Progress in Probability
and Statistics
Vol. 5

Edited by
Peter Huber
Murray Rosenblatt

Birkhäuser
Boston · Basel · Stuttgart

Seminar on Stochastic Processes, 1982

E. Çınlar,
K.L. Chung,
R.K. Getoor,
editors

1983

Birkhäuser
Boston • Basel • Stuttgart

Editors:

E. Çınlar
Technological Institute
Northwestern University
Evanston, Illinois 60201

K.L. Chung
Department of Mathematics
Stanford University
Stanford, California 94305

R.K. Getoor
Department of Mathematics
University of California - San Diego
La Jolla, California 92093

Library of Congress Cataloging in Publication Data

Seminar on Stochastic Processes (2nd : 1982 : Northwestern University)
Seminar on Stochastic Processes, 1982.

(Progress in probability and statistics ; v. 5)
1.Stochastic processes. I. Çınlar, E. (Erhan), 1941- . II. Chung, Kai Lai, 1917- . III. Getoor, R. K. (Ronald Kay), 1929- . IV. Title. V. Series.
QA274.A1S45 1982 519.2 83-15878

ISBN-13:978-0-8176-3131-4 **e-ISBN-13:978-1-4684-0540-8**
DOI: 10.1007/978-1-4684-0540-8

CIP-Kurztitelaufnahme der Deutschen Bibliothek

Seminar on Stochastic Processes:
Seminar on Stochastic Processes ... - Boston ; Basel ;
Stuttgart : Birkhäuser

2. 1982 (1983).
(Progress in probability and statistics ; Vol. 5)

NE: GT

Softcover reprint of the hardcover 1st edition 1983

TABLE OF CONTENTS

FOREWORD

This volume consists of about half of the papers presented during a three-day seminar on stochastic processes held at Northwestern University in March 1982. This was the second of such yearly seminars aimed at bringing together a small group of researchers to discuss their current work in an informal atmosphere.

The invited participants in this year's seminar were B. ATKINSON, R. BASS, K. BICHTELER, D. BURKHOLDER, K.L. CHUNG, J.L. DOOB, C. DOLEANS-DADE, H. FÖLLMER, R.K. GETOOR, J. GLOVER, J. MITRO, D. MONRAD, E. PERKINS, J. PITMAN, Z. POP-STOJANOVIC, M.J. SHARPE, and J. WALSH. We thank them and the other participants for the lively atmosphere of the seminar. As mentioned above, the present volume is only a fragment of the work discussed at the seminar, the other work having been committed to other publications.

The seminar was made possible through the enlightened support of the Air Force Office of Scientific Research, Grant No. 80-0252A. We are grateful to them as well as the publisher, Birkhäuser, Boston, for their support and encouragement.

E.Ç.
Evanston, 1983

Seminar on Stochastic Processes, 1982
Birkhäuser, Boston, 1983

GERM FIELDS AND A CONVERSE TO THE STRONG MARKOV PROPERTY

by

BRUCE W. ATKINSON

1. Introduction

The purpose of this paper is to give an intrinsic characterization of optional (i.e., stopping) times for the general germ Markov process, which includes the general right process as a special case. We proceed from the general to the specific.

In Section 2, we consider a "process" (G_t), which is a family of σ-fields, with the *germ Markov* property: the past and future are conditionally independent (c.i.) given the "germ" at the present (this includes the present and, in general, much more); see Section 2 for details. Knight [7] contains a discussion of such germs. For a random time R that is optional relative to (G_t), we show that, appropriately interpreted, the past before R and the future after R are c.i. given the germ at R, and moreover, that there is a very specific way of computing the expectation of a future event, relative to R, given the germ at R. The main tools involved are the projections of processes in the two time directions, most of which commute.

In Section 3, we take the Markov property at R described above, apply the projections, and prove that this property is sufficient for the optionality of R.

Finally, in Section 4, we interpret our results for processes (X_t) with right continuous paths. It turns out that the various fields associated to a random time R (done abstractly in Section 1) have nice interpretations in terms of the paths of (X_t). In general, the germ field of interest that replaces the present allows a "peek" into the future after R. For a right process, this germ field is generated simply by X_R and R, and we obtain the following pathwise description of optional times. R is optional if and only if the following holds: the pre-R process and the post-R process are c.i. given (X_R,R) on $\{R < \infty\}$, and given that $X_R = x$ and $R = r < \infty$, the post-R process has the same law as (X_t) starting at x. This is essentially the same result as that obtained in [9], although the method here is quite different.

2. Germ Markov Processes and the Strong Markov Property of Optional Times

In this section we review briefly some terminology and results from [1] and relate them to germ Markov processes.

We begin with a probability space (Ω, F, P) and a family $(G_t: 0 \le t \le \infty)$ of sub-σ-fields of F. We assume that G_t contains the set of all P-null sets relative to F for every t, and that $F = \vee_t G_t$. Of course, there is nothing essential about allowing $t = \infty$, but we do so for symmetry of definitions. For each Borel subset B of $[0,\infty]$, we let

$$F(B) \equiv \sigma(G_u: u \in B);$$

when B is an interval we shall write FB instead of $F(B)$.

(2.1) DEFINITION.

(a) *The strict past before* t:

$$F_{t-} \equiv F[0,t) \text{ for } t > 0, \quad F_{0-} \equiv \bigcap_{t>0} F_{t-}.$$

(b) *The past before* t:

$$F_{t} \equiv \bigcap_{s>t} F_{s-} \text{ for } t < \infty, \quad F_{\infty} \equiv F.$$

(c) *The strict future after* t:

$$\hat{F}_{t+} \equiv F(t,\infty] \text{ for } t < \infty, \quad \hat{F}_{\infty+} \equiv \bigcap_{t<\infty} \hat{F}_{t+}.$$

(d) *The future after* t:

$$\hat{F}_{t} = \bigcap_{s<t} \hat{F}_{s+} \text{ for } t > 0, \quad \hat{F}_{0} \equiv F.$$

(e) *The germ at* t:

$$F_{[t]} \equiv \bigcap_{r<t<s} F[r,s], \quad F_{[0]} \equiv F_{0}, \quad F_{[\infty]} \equiv \hat{F}_{\infty}.$$

(f) *The left germ at* t:

$$F_{[t)} \equiv \bigcap_{s<t} F(s,t) \text{ for } t > 0, \quad F_{[0)} \equiv F_{[0]}.$$

(g) *The right germ at* t:

$$F_{(t]} \equiv \bigcap_{s>t} F(t,s) \text{ for } t < \infty, \quad F_{(\infty]} \equiv F_{[\infty]}.$$

(2.2) DEFINITION. The family (G_t) is said to have

(a) the *germ Markov property* if, for every t, F_t and $\hat{F}_t$ are conditionally independent (which we abbreviate as c.i.) given $F_{[t]}$;

(b) the *left germ Markov property* if F_{t-} and $\hat{F}_t$ are c.i. given $F_{[t)}$ for every t;

(c) the *right germ Markov property* if F_t and $\hat{F}_{t+}$ are c.i. given $F_{(t]}$ for every t.

(2.3) THEOREM. *The three germ Markov properties of* (2.2) *are equivalent.*

PROOF. By symmetry, it is enough to check the equivalence of (2.2a) and (2.2b).

Suppose (2.2a). Let $0 < a < b < \infty$, $f \in bF_{a-}$, $g \in bF[a,b]$, and $h \in b\hat{F}_{b+}$. Then,

$$E[fgh \mid F_b \mid \hat{F}_a] = E[fg\ E[h \mid F_{[b]}] \mid \hat{F}_a] = E[f \mid F_{[a]}]g\ E[h \mid F_{[b]}].$$

Similarly,

$$E[fgh \mid \hat{F}_a \mid F_b] \equiv E[f \mid F_{[a]}]g\ E[h \mid F_{[b]}],$$

which is in $F(a-,b+) = \bigcap_{s<a<b<t} F[s,t]$. It follows from a monotone class argument that, for every $f \in bF$,

$$E[f \mid F_b \mid \hat{F}_a] = E[f \mid \hat{F}_a \mid F_b] = E[f \mid F(a-,b+)].$$

Fix $t > 0$, $s_n < t$, where $s_n \to t$. For each $0 < u < t$, apply the preceding result to the intervals (u,s_n) with large enough n, and let

$n \to \infty$. This gives, for every $f \in b\mathcal{F}$,

$$E[f \mid \mathcal{F}_{t-} \mid \hat{\mathcal{F}}_u] = E[f \mid \hat{\mathcal{F}}_u \mid \mathcal{F}_{t-}] = E[f \mid \mathcal{F}(u-,t)],$$

where $\mathcal{F}(u-,t) \equiv \bigcap_{s<u} \mathcal{F}[s,t)$. Now replace u by s_n and let $n \to \infty$ to get that, for every $f \in b\mathcal{F}$,

$$E[f \mid \mathcal{F}_{t-} \mid \hat{\mathcal{F}}_t] = E[f \mid \hat{\mathcal{F}}_t \mid \mathcal{F}_{t-}] = E[f \mid \mathcal{F}_{[t)}],$$

which is equivalent to (2.2b).

The proof of that (2.2b) implies (2.2a) is similar. □

(2.4) REMARK. Compare this with [7], Theorem 1, where a similar equivalence is shown for slightly different definitions of the past, future, and germ.

Due to the equivalence of the germ and right germ Markov properties, there are two ways to formulate the strong Markov property at an optional time. Before doing so, we shall recall the main results of [1].

(2.5) DEFINITION. $\mathcal{M}$ (resp. $\mathcal{O}$, $\mathcal{P}$, $\hat{\mathcal{O}}$, $\hat{\mathcal{P}}$) is the σ-field on $[0,\infty] \times \Omega$ generated by all real valued processes Z such that Z_t is measurable relative to $\mathcal{F}$ (resp. $\mathcal{F}_t$, $\mathcal{F}_{t-}$, $\hat{\mathcal{F}}_t$, $\hat{\mathcal{F}}_{t+}$) for every t and $t \to Z_t$ is right (resp. right, left, left, right) continuous P- almost surely.

For each $Z \in b\mathcal{M}$, we define ${}^oZ$ (resp. ${}^pZ$, ${}^{\hat{o}}Z$, ${}^{\hat{p}}Z$) to be the optional (resp. predictable, co-optional, co-predictable) projection of Z relative to the family $(\mathcal{G}_t)$; see [1], Section 3. In other words, the mapping $Z \to {}^oZ$ (resp. ${}^pZ$, ${}^{\hat{o}}Z$, ${}^{\hat{p}}Z$) preserves right (resp. left, left, right) continuity, and for each t, ${}^{o}Z_t$ (resp. ${}^{p}Z_t$, ${}^{\hat{o}}Z_t$, ${}^{\hat{p}}Z_t$) is the conditional expectation of Z_t given F_t (resp. F_{t-}, $\hat{F}_t$, $\hat{F}_{t+}$).

For the rest of this section, we assume (G_t) is germ Markov. It follows from (2.2) and (2.3) that ${}^{o\hat{o}}Z$ (resp. ${}^{o\hat{p}}Z$, ${}^{p\hat{o}}Z$) is a modification of ${}^{\hat{o}o}Z$ (resp. ${}^{\hat{p}o}Z$, ${}^{\hat{o}p}Z$) where, e.g., ${}^{o\hat{o}}Z \equiv {}^{o}({}^{\hat{o}}Z)$. It is shown in [1], Section 3, that ${}^{o\hat{o}}Z$ (resp. ${}^{o\hat{p}}Z$, ${}^{p\hat{o}}Z$) and ${}^{\hat{o}o}Z$ (resp. ${}^{\hat{p}o}Z$, ${}^{\hat{o}p}Z$) are indistinguishable.

(2.6) REMARK. It is not difficult to show that, for any family (G_t) (not necessarily germ Markov), the above commutativity of projections is equivalent to having

$$E[f \mid F_t \mid \hat{F}_t] = E[f \mid \hat{F}_t \mid F_t]$$

for all t and all $f \in bF$. If we stipulate further that $F_{[t]} = F_t \cap \hat{F}_t$ for all t, then the commutativity of projections is equivalent to the germ Markov property.

(2.7) DEFINITION. A mapping $R: \Omega \to [0,\infty]$ is called a random time if R is F-measurable. A random time R is said to be optional if $\{R < t\} \in F_t$ for every t; it is said to be *predictable* if it is optional and $\{(t,\omega): t = R(\omega)\} \in P$.

It follows from (2.2) that, for every t,

$$F_{[t]} = F_t \cap \hat{F}_t,\quad F_{[t)} = F_{t-} \cap \hat{F}_t,\quad F_{(t]} = F_t \cap \hat{F}_{t+}.$$

Thus the following definitions of past, future, and germ fields at a random time R are consistent with our notations:

(2.8)
$$\begin{aligned} F_R &= \sigma(Z_R\colon Z \in b\mathcal{O}), \quad F_{R-} = \sigma(Z_R\colon Z \in b\mathcal{P}), \\ \hat{F}_R &= \sigma(Z_R\colon Z \in b\hat{\mathcal{O}}), \quad \hat{F}_{R+} = \sigma(Z_R\colon Z \in b\hat{\mathcal{P}}), \\ F_{[R]} &= \sigma(Z_R\colon Z \in b(\mathcal{O} \cap \hat{\mathcal{O}})), \\ F_{[R)} &= \sigma(Z_R\colon Z \in b(\mathcal{P} \cap \hat{\mathcal{O}})), \\ F_{(R]} &= \sigma(Z_R\colon Z \in b(\mathcal{O} \cap \hat{\mathcal{P}})). \end{aligned}$$

We might naturally call $F_{[R]}$, $F_{[R)}$, $F_{(R]}$ the germ, left germ, right germ at R.

The next result verifies two different strong Markov properties, (SM_1) and (SM_2), for optional times and a predictable strong Markov property (PSM) for predictable times. These correspond to the three equivalent formulations of the germ Markov property; see (2.2) and (2.3).

(2.9) THEOREM. *Let* R *be an optional time. Then* (SM_1) *and* (SM_2) *hold. If* R *is predictable, then* (PSM) *holds.*

(SM_1) F_R, $\hat{F}_R$ *are c.i. given* $F_{[R]}$; $E(Z_R) = E({}^oZ_R)$ $\forall\, Z \in b\hat{\mathcal{O}}$.

(SM_2) F_R, $\hat{F}_{R+}$ *are c.i. given* $F_{(R]}$; $E(Z_R) = E({}^oZ_R)$ $\forall\, Z \in b\hat{\mathcal{P}}$.

(PSM) F_{R-}, $\hat{F}_R$ *are c.i. given* $F_{[R)}$; $E(Z_R) = E({}^pZ_R)$ $\forall\, Z \in b\hat{\mathcal{O}}$.

PROOF. Suppose R is optional and $Z \in b\hat{\mathcal{O}}$. Then, $E(Z_R \mid F_R) = {}^oZ_R$. Moreover, ${}^oZ \in b(\mathcal{O} \cap \hat{\mathcal{O}})$ by commutativity of the optional and co-optional projections. This implies that ${}^oZ_R \in bF_{[R]}$. Hence, (SM_1) holds. The property (SM_2) follows similarly by the commutativity of the optional and co-predictable projections. Finally, similar reasoning shows (PSM) for R predictable by using commutativity of the predictable and co-optional projections. □

In the next section, we prove that (SM_1) and (SM_2) separately provide an intrinsic characterization of optional times for a germ Markov family (G_t), and (PSM) provides such a characterization for predictable times. In Section 4, we assume the fields (G_t) are generated by a process (X_t) with right continuous paths. The various fields involved have particularly nice interpretations in terms of the paths of X. If X is a right process, then (SM_2) can be thought of strictly in terms of the pre-R and post-R processes to obtain a pathwise characterization of optional times.

(2.10) REMARK. It is clear that one can also analyse reverse optional and reverse predictable times by reversing time in all our arguments. In particular, it can be shown that, if R is reverse predictable, then F_R and $\hat{F}_{R+}$ are c.i. given $F_{(R]}$. We shall use this remark in Section 4; see (4.12).

3. Converse to the Strong Markov Property

Here we assume that (G_t) has the germ Markov property (2.2). The key to the proofs of this section is the fact that we may "split" the σ-field M of measurable processes into processes adapted to the past and processes adapted to the future. This is obviously analogous to the fact that $F = F_t \vee \hat{F}_t$ for every t (compare with [4], Lemma 2.1).

(3.1) LEMMA. $M = P \vee (O \cap \hat{O}) \vee \hat{P}$.

PROOF. By the definition of M, it is enough to check that $Z \in b(P \vee (O \cap \hat{O}) \vee \hat{P})$ where $Z_t(\omega) = F(\omega)$, $F \in bG_u$, u fixed. Write $Z_t = U_t + V_t + W_t$ where

$$U_t = F\, 1_{[0,u)}(t), \quad V_t = F\, 1_{\{u\}}(t), \quad W_t = F\, 1_{(u,\infty]}(t).$$

Since $U \in b\hat{P}$, $V \in b(O \cap \hat{O})$, $W \in bP$, the result obtains. □

(3.2) COROLLARY. *Let* R *be a random time. Then,*

(a) $F = F_{R-} \vee F_{[R]} \vee \hat{F}_{R+}$;

(b) $O = P \vee (O \cap \hat{O})$ *and* $F_R = F_{R-} \vee F_{[R]}$;

(c) $\hat{O} = (O \cap \hat{O}) \vee \hat{P}$ *and* $\hat{F}_R = F_{[R]} \vee \hat{F}_{R+}$.

PROOF. (a) This follows immediately from (3.1).

(b) Let $U \in bP$, $V \in b(O \cap \hat{O})$, $W \in b\hat{P}$. Then, ${}^{o}(UVW) = U \cdot V \cdot {}^{o}W \in b(P \vee (O \cap \hat{O}))$ since ${}^{o}W \in b(O \cap \hat{P}) \subset b(O \cap \hat{O})$. Thus, (3.1) implies that $O = P \vee (O \cap \hat{O})$. This in turn implies that $F_R = F_{R-} \vee F_{[R]}$.

(c) Similar to the proof of (b). □

Before proceeding, here are a few comments on random measures; see [1]. In this section, all random measures A(dt) are *finite* in the sense that $E\, A([0,\infty]) < \infty$. For such a random measure A, the dual *optional projection* of A, denoted by A^o, is the random measure defined by

$$E \int Z_t\, A^o(dt) = E \int {}^{o}Z_t\, A(dt) \quad \forall\, Z \in bM.$$

Dual predictable, dual co-optional, and *dual co-predictable projections* of A, denoted by A^p, $A^{\hat{o}}$, $A^{\hat{p}}$, respectively, are defined similarly. For a finite measure A, we will write $A \in O$ (resp. P, $\hat{O}$, $\hat{P}$) to mean that $A = A^o$ (resp. A^p, $A^{\hat{o}}$, $A^{\hat{p}}$), and in this case, we call A optional (resp. predictable, co-optional, co-predictable). Finally, if $Z \in bM^+$, then

$Z * A$ will denote the finite random measure $Z * A(dt) = Z_t\, A(dt)$.

The following gives an intrinsic characterization of optional and predictable random measures. This is quite similar to a result of Sharpe [10] where raw additive functionals are characterized.

(3.3) THEOREM. a) $A \in \mathcal{O}$ *if and only if* $(Y * A)^{\hat{o}} \in \mathcal{O} \cap \hat{\mathcal{O}}$ *for every* $Y \in b\mathcal{O}^+$.

b) $A \in \mathcal{O}$ *if and only if* $(Y * A)^{\hat{p}} \in \mathcal{O} \cap \hat{\mathcal{P}}$ *for every* $Y \in b\mathcal{O}^+$.

c) $A \in \mathcal{P}$ *if and only if* $(Y * A)^{\hat{o}} \in \mathcal{P} \cap \hat{\mathcal{O}}$ *for every* $Y \in b\mathcal{P}^+$.

PROOF. We show only (a), as (b) and (c) are quite similar. If $A \in \mathcal{O}$, then $Y * A \in \mathcal{O}$ for every $Y \in b\mathcal{O}^+$. By commutativity of dual projections, which follows from that of projections of processes, we have $(Y * A)^{\hat{o}} \in \mathcal{O} \cap \hat{\mathcal{O}}$.

Conversely, suppose $(Y * A)^{\hat{o}} \in \mathcal{O} \cap \hat{\mathcal{O}}$ for every $Y \in b\mathcal{O}^+$. Then, for $Y \in b\mathcal{O}^+$ and $Z \in b\hat{\mathcal{O}}^+$,

$$
\begin{aligned}
E \int Y_t\, Z_t\, A(dt) &= E \int Z_t\, (Y * A)^{\hat{o}}\,(dt) \\
&= E \int {}^{o}Z_t\, (Y * A)^{\hat{o}}\,(dt) \\
&= E \int {}^{o}Z_t\, Y_t\, A(dt) = E \int {}^{o}(YZ)_t\, A(dt)
\end{aligned}
$$

where the third equality is justified since ${}^{o}Z \in b(\mathcal{O} \cap \hat{\mathcal{O}})^+$. By (3.1), $\mathcal{M} = \mathcal{O} \vee \hat{\mathcal{O}}$, which is generated by processes of form $Y \cdot Z$ with Y and Z as above. So, for every $W \in b\mathcal{M}$,

$$
E \int W_t\, A(dt) = E \int {}^{o}W_t\, A(dt),
$$

in other words, $A = A^{o}$. □

(3.4) REMARKS. a) Of course, there are similar characterizations of co-optional and co-predictable random measures.

b) Using (3.1) again, it can be shown that $A \in \mathcal{O}$ if and only if $(Y * A)^o = {}^oY * A^o$ for every $Y \in b\hat{\mathcal{O}}^+$.

We now apply (3.3) to the special case where $A = \varepsilon_R$, unit mass at random time R.

(3.5) THEOREM. *Let* R *be a random time. If either* (SM_1) *or* (SM_2) *holds, then* R *is optional. If* (PSM) *holds, then* R *is predictable.*

PROOF. Suppose (SM_1) holds, and let $Y \in b\mathcal{O}^+$ and $Z \in b\hat{\mathcal{O}}^+$. Choose $\bar{Y} \in b(\mathcal{O} \cap \hat{\mathcal{O}})^+$ so that $E[Y_R \mid \hat{F}_R] = \bar{Y}_R$. Then,

$$E \int Z_t \; (Y * \varepsilon_R)\,(dt) = E[Z_R\, Y_R] = E[Z_R\, \bar{Y}_R]$$

$$= E[{}^oZ_R\, \bar{Y}_R] = E[{}^oZ_R\, Y_R]$$

$$= E \int Z_t \; (Y * \varepsilon_R)^o\,(dt).$$

As $Z \in b\hat{\mathcal{O}}^+$ is arbitrary, $(Y * \varepsilon_R)^{\hat{o}} = (Y * \varepsilon_R)^{o\hat{o}} \in \mathcal{O} \cap \hat{\mathcal{O}}$. By (3.3), ε_R is optional; hence R is optional.

The proof that (SM_2) implies the optionality of R is similar, using (3.3b). The proof that (PSM) implies the predictability of R uses (3.3c) and [2], V-T26, IV-T15. □

4. Applications to Processes with Right Continuous Paths

Let E be a U-space, a universally measurable subset of a compact metric space, and let $\mathcal{E}$ be its set of Borel subsets. Let Ω be the space of right continuous paths $\omega: [0,\infty) \to E$, and let $X_t(\omega) = \omega(t)$ for each t.

(4.1) DEFINITIONS.

a) $G_t^o \equiv \sigma(X_t)$ for $t < \infty$, $G_\infty^o \equiv \bigcap_{t<\infty} \sigma(X_s;\ s \geq t)$, $F^o \equiv \vee_t\ G_t^o$.

b) F_t^o, F_{t-}^o, $\hat{F}_t^o$, $\hat{F}_{t+}^o$, $F_{[t]}^o$, $F_{[t)}^o$, $F_{(t]}^o$ are defined as in (2.1), but relative to the family (G_t^o).

c) M^o, O^o, P^o, $\hat{O}^o$, $\hat{P}^o$ are defined as in (2.5), relative to the family (G_t^o), with the phrase, "P a.s." omitted.

d) For $t < \infty$, we define θ_t, a_t, $b_t: \Omega \to \Omega$ by $\theta_t\omega(s) = \omega(t+s)$, $a_t\omega(s) = \omega(t \wedge s)$, $b_t\omega(s) = \omega(t \vee s)$.

The next result does not assume any probability measure on Ω.

(4.2) THEOREM. *On* $R^+ \times \Omega$, *let* S_1 *be the σ-field generated by the mapping* $(t,\omega) \to (t, b_t\omega)$ *and* S_2 *that by* $(t,\omega) \to (t,\theta_t\omega)$. $S_1 = S_2 = \hat{P}^o$ restricted to $R^+ \times \Omega$.

PROOF. Right continuity of paths and a monotone class argument imply that, for every $t < \infty$, and $F \in bF^o$, $F \in b\hat{F}_{t+}^o$ if and only if $F \circ b_t = F$.

Let Y be right continuous and adapted to $(\hat{F}_{t+}^o)$. Then, $Y_t = Y_t \circ b_t$. Hence, $Y \in bS_1$. So, $\hat{P}^o \subset S_1$. On the other hand, if $Y_t \equiv g(t)\, f(X_{t\vee s})$ for $t < \infty$ and $s < \infty$ for some $g \in bC(R^+)$ and $f \in bC(E)$ (i.e. bounded and continuous on R^+ and on E respectively), then clearly, $Y \in b\hat{P}^o$. By the monotone class theorem, we conclude that $\hat{P}^o \supset S_1$. Hence, $\hat{P}^o = S_1$.

A similar reasoning shows that $\hat{P}^o \supset S_2$. There remains to prove that $S_1 \subset S_2$. For this, it is enough to check that $Y \in bS_2$ if Y has the form $Y_t = f(X_{t\vee s})$ where $s < \infty$ and $f \in bC(E)$ are fixed. Using the continuity of f and a dyadic approximation procedure, this reduces to checking that $f(X_c)\, 1_{[a,b)}(t) \in bS_2$ whenever $a < b \le c$. Then, we write

$$f(X_c)\, 1_{[a,b)}(t) = \lim_n \sum_k f(X(c - k_n + t)\, 1_{[k_n,(k+1)_n)}(t)$$

where $k_n \equiv k2^{-n}$ and the summation is over all k such that $k_n \le a < (k+1)_n$ or $a \le k_n < (k+1)_n < b$. Now, each summand is in bS_2 and the conclusion follows. The proof is complete. □

We now assume that (Ω, F^o) is equipped with a probability P. We let $G_t \equiv G_t^o \vee$ (P-null sets w.r.t. F^o). We shall assume that (G_t) is germ Markov relative to P.

Let I denote the collection of all evanescent processes, that is, $I = \{Z\colon P(Z_t \ne 0$ for some $t) = 0\}$. Then, it follows that, relative to (G_t), we have $M = M^o \vee I$, $O = O^o \vee I$, $P = P^o \vee I$, $\hat{O} = \hat{O}^o \vee I$, $\hat{P} = \hat{P}^o \vee I$. We now improve (3.1) under our hypothesis on (G_t).

(4.3) THEOREM. *With* (G_t) *as above*, $M = P \vee \hat{P}$.

PROOF. By the monotone class theorem, we need only check that $Y \in b(P \vee \hat{P})$ if Y has the form $Y_t(\omega) = f(X_s(\omega))$ with $f \in bC(E)$ and $s < \infty$ fixed. As in (3.1), we write $Y_t = U_t + V_t + W_t$ where $U_t = f(X_s)\, 1_{[0,s)}(t)$, $V_t = f(X_s)\, 1_{\{s\}}(t)$, $W_t = f(X_s)\, 1_{(s,\infty]}(t)$. Then, as in (3.1), $U \in b\hat{P}$ and $W \in bP$. By right continuity of $t \to f(X_t)$, $V \in b\hat{P}$, and we are finished. □

Theorem (4.2) tells us exactly what $\hat{P}$ is in terms of paths. We now focus on P.

(4.4) LEMMA. *Let* S_3 *be the σ-field on* $R^+ \times \Omega$ *generated by the mapping* $(t,\omega) \to (t, a_t\omega)$. *Then, on* $R^{++} \times \Omega$, *we have* $P \subset S_3 \vee I$. ($R^{++} \equiv (0,\infty)$.).

PROOF. Let (Y_t) be bounded, left-continuous in t, and adapted to (F^o_{t-}). By dyadic approximation, it is enough to check the case where $Y_t = f(X_c)\, 1_{(a,b]}(t)$ where $0 < c < a < b$ and $f \in bC(E)$. But, then, $Y_t(\omega) = Y_t(a_t\omega)$, which is what was needed. □

Combining (4.2), (4.3), and (4.4) we have:

(4.5) THEOREM. *On* $R^+ \times \Omega$, M *is generated by* I *and the mapping* $(t,\omega) \to (t, a_t\omega, \theta_t\omega)$.

(4.6) REMARK. In [3], IV. 97, it is shown that, if Ω is the space of right continuous paths with left-limits in E, then the mapping $(t,\omega) \to (t, a_t\omega)$ generates the optional σ-field relative to the natural uncompleted and unregularized filtration $(\sigma(G^o_s : s \le t))$. The mapping $(t,\omega) \to (t, \theta_t\omega)$ is mentioned there but not studied. Also, (4.5) is a slightly different version of [11], (26.1), for right processes.

So far in this section we made no use of the assumption that (G_t) is germ Markov. We now do so. Compare [6], Section 2.

(4.7) THEOREM. *Let* (G_t) *be germ Markov and* R *a random time. Then,*

$$F_{[R]} = F_{[R)} \vee F_{(R]}, \quad F_R = F_{R-} \vee F_{(R]}, \quad \hat{F}_R = F_{[R)} \vee \hat{F}_{R+}.$$

PROOF. Let $Y \in b\mathcal{P}$ and $Z \in b\hat{\mathcal{P}}$. Then, ${}^{o\hat{o}}(YZ) = {}^{\hat{o}}Y\,{}^{o}Z \in (\mathcal{P} \cap \hat{\mathcal{O}}) \vee (\mathcal{O} \cap \hat{\mathcal{P}})$ by the commutativity of projections. By (4.3) and a monotone class argument, $\mathcal{O} \cap \hat{\mathcal{O}} = (\mathcal{P} \cap \hat{\mathcal{O}}) \vee (\mathcal{O} \cap \hat{\mathcal{P}})$. Thus, by definition, $\mathcal{F}_{[R]} = \mathcal{F}_{[R)} \vee \mathcal{F}_{(R]}$. The other assertions follow by similar reasoning. □

We next study the various fields associated with R in terms of the paths of the process. First some definitions: Let R be a random time; on $\{R < \infty\}$, define

(4.8) a) $G(R) \equiv \bigcap_{t>0} \sigma(R;\ X_{R+s},\ s \le t;\ \text{P-null sets}),$

b) $M(R) \equiv \sigma(X_{R\wedge t},\ t \ge 0;\ \text{P-null sets}),$

c) $N(R) \equiv \sigma(X_{R+t},\ t \ge 0;\ \text{P-null sets}).$

(4.9) THEOREM. (a) *On* $\{R < \infty\}$, $\mathcal{F}_{(R]} \subset G(R) \subset \mathcal{F}_{R+}$.

(b) *On* $\{R < \infty\}$, $\mathcal{F}_R = M(R) \vee \mathcal{F}_{(R]} \vee \mathcal{F}_0$.

(c) *On* $\{R < \infty\}$, $\hat{\mathcal{F}}_{R+} = N(R) \vee \sigma(R)$.

(d) *Suppose* $\mathcal{F}_R$ *and* $\hat{\mathcal{F}}_{R+}$ *are c.i. given* $\mathcal{F}_{(R]}$. *Then,* $\mathcal{F}_R$ *and* N(R) *are c.i. given* G(R) *on* $\{R < \infty\}$.

PROOF. Let $Y \in b\mathcal{M}$ be right continuous. Then,

$$ {}^{o\hat{p}}Y_t = F\,1_{\{0\}}(t) + Z_t\,1_{(0,\infty)}(t), \quad t < \infty, $$

where $F \in b\mathcal{F}_{(0]}$ and (Z_t) is a right continuous version of $E[Y_t \mid \mathcal{F}_{(t]}]$. This is a consequence of the right germ Markov property (2.2c) and the fact that the optional and co-predictable projections preserve right continuity. Clearly, $F\,1_{\{R=0\}} \in bG(R)$. We now show that $Z_R\,1_{\{0<R<\infty\}} \in bG(R)$.

Let $t > 0$. By dyadic approximation, $Z_R 1_{\{0 < R < \infty\}}$ can be written as a limit of sums of terms of the form $F 1_{\{a \le R < b\}}$, where $0 < a < b < a + t$ and $F \in b\mathcal{F}_{(b]}$. Now, if $b < u < a + t$ and $f \in bC(E)$, then

$$f(X_u) 1_{\{a \le R < b\}} = \lim_n \sum_n f(X_{R+u-k_n}) 1_{\{k_n \le R < (k+1)_n\}}$$

where $k_n = k2^{-n}$ and the summation is over k such that $k_n \le a < (k+1)_n$ or $a \le k_n < (k+1)_n < b$. For n large enough, $0 < u - k_n < t$, and hence

$$f(X_u) 1_{\{a \le R < b\}} \in G(R,t) \equiv \sigma(R;\ X_{R+s},\ s \le t;\ P\text{-null sets}).$$

It follows by a monotone class argument that, if $G \in b\sigma(X_u,\ b < u < a + t$; P-null sets), then $G 1_{\{a \le R < b\}} \in b\, G(R,t)$. Thus, if $F \in b\mathcal{F}_{(b]}$ then $F 1_{\{a \le R < b\}} \in b\, G(R,t)$. In summary, $Z_R 1_{\{0 < R < \infty\}} \in b\, G(R,t)$ for every $t > 0$; that is, $Z_R 1_{\{0 < R < \infty\}} \in b\, G(R)$.

We now have that $\mathcal{F}_{(R]} \subset G(R)$ on $\{R < \infty\}$. By (4.2) it follows that $G(R,t) \subset \hat{\mathcal{F}}_{R+}$ for every $t > 0$. Hence, $G(R) \subset \hat{\mathcal{F}}_{R+}$ on $\{R < \infty\}$, and (a) is proved.

By (4.4), $\mathcal{F}_{R-} \subset M(R) \vee \mathcal{F}_{(R]} \vee \mathcal{F}_0$ on $\{R < \infty\}$. Now, (b) follows from (4.7); (c) follows directly from (4.2).

Suppose $\mathcal{F}_R$ and $\hat{\mathcal{F}}_{R+}$ are c.i. given $\mathcal{F}_{(R]}$. Certainly, the same holds on $\{R < \infty\}$. Since $\mathcal{F}_{(R]} \subset G(R) \subset \hat{\mathcal{F}}_{R+} = \hat{\mathcal{F}}_{R+} \vee \mathcal{F}_{(R]}$ on $\{R < \infty\}$, by [7], Lemma 1(a), we have that $\mathcal{F}_R$ and $\hat{\mathcal{F}}_{R+}$ are c.i. given $G(R)$ on $\{R < \infty\}$. Since $N(R) \subset \hat{\mathcal{F}}_{R+}$ on $\{R < \infty\}$, this implies (d). □

(4.10) REMARKS. (a) Obviously, $G(R)$ has a natural interpretation as a right germ field at R. We emphasize here that the definition of $G(R)$ involves first adjoining R then taking intersections over $t > 0$.

There is nothing, a priori, to suggest that this would be the same as performing these operations in the opposite order. (In fact, in [8], §2b, there is an example where the two orders yield different fields.) In view of (2.10), if R is reconstructable, e.g. a last hitting time of certain sets, then R splits the process X into two independent pieces given $G(R)$, the right germ field. This is an extension, to germ Markov processes, of the fact, proved in [5], that such a splitting occurs for a right process at co-optional times given $X(R)$. See also (4.12).

Note also that if, in fact, (d) obtains and $G(R) \subset F_R$ on $\{R < \infty\}$, then $G(R) = F_R \cap \hat{F}_{R+} = F_{(R]}$ on $\{R < \infty\}$.

(b) Another nice property of $F_{(R]}$ is that it is countably generated (modulo null sets) which is not clear for $G(R)$. Indeed, by right continuity of paths, M^o is countably generated. But since, modulo evanescent sets, $b(O \cap \hat{P}) = \{^{o\hat{p}}Z\colon Z \in bM^o\}$, it is clear that, modulo evanescent sets, $O \cap \hat{P}$ is countably generated and this implies our assertion about $F_{(R]}$.

For the rest of this section, we shall assume that (X_t) is a right process (see [11] for definitions and development) whose transition function (P_t) is Borel, i.e. $P_t\colon bE \to bE$. We also assume that P is of the form P^μ for some fixed initial law μ. Now $F_{(R]}$ has a particularly tractable form.

(4.11) THEOREM. *On* $\{R < \infty\}$, *let* $S = \sigma(R, X_R$, *P-null sets*). *Then,* $F_{(R]} = S$ *on* $\{R < \infty\}$.

PROOF. It follows from the right continuity of paths that $S \subset F_{(R]}$. Next, let $Y_t = f(t)\, F \circ \theta_t$ for $t < \infty$, where $f \in bB(R^+)$

and $F \in bF^o$. By the strong Markov property, ${}^oY_t = f(t)\, E^{X(t)}F$ for $t < \infty$, and hence

$$ {}^oY_R \, 1_{\{R<\infty\}} = f(R)\,(E^{X(R)}F)\, 1_{\{R<\infty\}} \in bS $$

(here we used the fact that P_t: $bE \to bE$). By the monotone class theorem, the fact that $b(\mathcal{O} \cap \hat{\mathcal{P}}) = \{{}^oY: Y \in b\hat{\mathcal{P}}\}$, and (4.2), it follows that $F_{(R]}$ restricted to $\{R < \infty\}$ is contained in S, and the proof is complete. □

(4.12) REMARK. Combining (2.10) and (4.11) we have the following: suppose R is a reverse predictable time (such an R would be called reconstructable by Dynkin, Getoor, and Sharpe). It follows that on $\{R < \infty\}$, F_R and $\hat{F}_{R+}$ are c.i. given X_R and R. Since (X_t) is a right process, then either by the strong Markov property in conjunction with (4.5), or by [3], IV.97, and the right continuity of the natural completed filtration, it follows that $F_R = M(R) \vee \sigma(R)$. (See (4.8b).) Similarly, by (4.2), $F_{R+} = N(R) \vee \sigma(R)$. (See (4.8c).) In words, we thus have the pre-R process together with R, and the post-R process together with R, are c.i. given X_R and R on $\{R < \infty\}$. This extends the Markov property at reconstructable co-optional times (a specific type of reverse predictable time) proved in [5]. The same result can, most likely, be handled by space-time techniques as described in [6].

Recall (SM_2) (see (2.9)). By (2.9) and (3.5), R is optional if and only if (SM_2) holds. In view of the above, this has a very nice interpretation for right processes. We use the fact that $F_{(R]} = \sigma(X_R$; R; P-null sets) and $F_R = M(R) \vee \sigma(R)$, $\hat{F}_{R+} = N(R) \vee \sigma(R)$, as remarked in (4.12).

(4.13) THEOREM. *Let* R *be a random time. Consider the following statements:*

(A) $M(R)$ *and* $N(R)$ *are c.i. given* X_R, R *on* $\{R < \infty\}$;

(B) *for every* $f \in b\mathcal{B}(R^+)$ *and every* $F \in b\mathcal{F}^o$,

$$E[f(R) \cdot F \circ \theta_R \cdot 1_{\{R<\infty\}}] = E[f(R)\ (E^{X(R)}F)\ 1_{\{R<\infty\}}];$$

(B') *for every* $F \in b\mathcal{F}^o$, *almost surely on* $\{R < \infty\}$,

$$E[F \circ \theta_R \mid X_R, R] = E^{X(R)}F.$$

Then, [R is optional] $\iff$ [(A) and (B)] $\iff$ [(A) and (B')].

PROOF. (A) is obviously equivalent to the statement that $M(R) \vee \sigma(R)$ and $N(R) \vee \sigma(R)$ are c.i. on $\{R < \infty\}$ given X_R and R. Thus, by (4.2), (2.9), and (3.5), [R is optional] $\iff$ [(A) and (B)]. Clearly, (B') $\Rightarrow$ (B). Now suppose R is optional. Then, if $f \in b\mathcal{E}$ and $g \in b\mathcal{B}(R^+)$, we have

$$\begin{aligned} E[f(X_R)\ g(R)\ F \circ \theta_R\ 1_{\{R<\infty\}}] &= E[^o(f(X_.)\ g(\cdot)\ F \circ \theta_.)_R\ 1_{\{R<\infty\}}] \\ &= E[f(X_R)\ g(R)(E^{X(R)}F)\ 1_{\{R<\infty\}}], \end{aligned}$$

which shows that (B') holds. □

Roughly speaking, the theorem states that R is optional if and only if the pre-R process and the post-R process are c.i. on $\{R < \infty\}$ given X_R and R, and moreover, given that $X_R = x$ and $R = r < \infty$, the post-R process has the same law as (X_t) starting at x. We note that the value of R (in condition (B')) is irrelevant in the computation of the transitions after R; this may be interpreted as the preservation of time homogeneity at the random time R. Thus, for right processes, we may replace the traditional definition of optional times with the more probabilistic one described above.

(4.14) REMARK. In the preceding theorem, we may replace (A) by

(A') $M(R) \vee \sigma(R)$ and $N(R)$ are c.i. given X_R on $\{R < \infty\}$.

In other words, in the presence of either (B) or (B'), (A) is equivalent to (A'). This follows from the observation that, for any $f \in bE$,

$$f(X_R)\, 1_{\{R<\infty\}} = \lim_n f(X_{R\wedge n})\, 1_{\{R<n\}} \in b(M(R) \vee \sigma(R)).$$

In particular, then, we have that R is optional if and only if (A') and (B') hold. It is this formulation that is found in [9], Corollary 4.3.

References

1. B. ATKINSON. Generalized strong Markov properties and applications. *Z. Wahrscheinlichkeitstheorie verw. Gebiete 60* (1982), 71-78.
2. C. DELLACHERIE. *Capacités et Processus Stochastiques.* Springer-Verlag, Berlin, 1972.
3. C. DELLACHERIE and P.A. MEYER. *Probabilities and Potential.* North-Holland, Amsterdam, 1978.
4. E.B. DYNKIN. Additive functionals of Markov processes and stochastic systems. *Ann. Inst. Fourier (Grenoble) 25,* 3 et 4 (1975), 177-200.
5. R.K. GETOOR and M.J. SHARPE. The Markov property at co-optional

times. *Z. Wahrscheinlichkeitstheorie verw. Gebiete 48* (1979), 201-211.

6. R.K. GETOOR and M.J. SHARPE. Markov properties of a Markov process. *Z. Wahrscheinlichkeitstheorie verw. Gebiete 55* (1981), 313-330.

7. F.B. KNIGHT. A remark on Markovian germ fields. *Z. Wahrscheinlichkeitstheorie verw. Gebiete 15* (1970), 291-296.

8. D. MALON. Germ Markov processes. Ph.D. Dissertation, Northwestern University, 1981.

9. A.O. PITTENGER. Regular birth times for Markov processes. *Ann. Probab. 9* (1981), 769-780.

10. M.J. SHARPE. Killing times for Markov processes. *Z. Wahrscheinlichkeitstheorie verw. Gebiete 58* (1981), 223-230.

11. M.J. SHARPE. *General Theory of Markov Processes* (to appear).

B. W. ATKINSON
Department of Mathematics
University of Southern California
Los Angeles, CA 90089

Current Address:

Department of Mathematics
University of Florida
Gainesville, FL 32611

Seminar on Stochastic Processes, 1982
Birkhäuser, Boston, 1983

APPLICATIONS OF REVUZ AND PALM TYPE MEASURES FOR ADDITIVE FUNCTIONALS IN WEAK DUALITY

by

B. W. ATKINSON and J. B. MITRO

0. Introduction

Several characterizations of additive functionals of a Markov process have been described in recent years. Under strong (Hunt) duality hypotheses this was accomplished in a series of papers by Revuz [14],[15], Getoor [9], and Sharpe [17]; for "symmetric" processes this was done by Fukushima [7] and Dynkin [4],[5]; earlier, the situation for Markov stochastic systems was investigated by Dynkin [3],[6]. Here, we obtain results along the same lines for processes in *weak* duality. The main tool is the "auxiliary process" [13] associated to a pair of Markov processes in weak duality. (Some facts about this process are recalled below.) Our approach is guided in part by similarities with the theory of flows ([8],[16]) and exploits the interplay between optionality and cooptionality in this context.

A pair of right processes (Ω, F, P^x, X) and $(\hat{\Omega}, \hat{F}, \hat{P}^x, \hat{X})$ with state space (E,E) are in *weak duality* relative to the σ-finite measure ξ if their resolvents satisfy

$$(0.1) \qquad \int f(x)\, U^\alpha g(x)\, \xi(dx) = \int f\hat{U}^\alpha(x)\, g(x)\, \xi(dx)$$

for $f, g \in bE^+$, $\alpha > 0$. This "switching" property holds with the potential operators replaced by the corresponding transition operators, and more generally, for $f_j \in bE^+$ $(0 \le j \le n)$ and $0 = t_0 \le t_1 \le \cdots \le t_n$,

$$E^\xi[\prod_{j=0}^{n} f_j(X_{t_j})] = \hat{E}^\xi[\prod_{j=0}^{n} f_{n-j}(\hat{X}_{t_j})], \tag{0.2}$$

where $E^\xi(\cdot) = \int E^x(\cdot)\,\xi(dx)$. The basis for the construction of the auxiliary process is a pair of Borel right Markov processes in weak duality, for which we assume in addition that all sample paths have left limits in E at all strictly positive times prior to the lifetime.

Denote by $\underset{\sim}{\Omega}$ the space of paths from $\mathbb{R}$ into $E \cup \Delta \cup \hat{\Delta}$ which admit a random birth time $\hat{\underset{\sim}{\zeta}}$ and a random death time $\underset{\sim}{\zeta}$ and which are right continuous with left limits on $(\hat{\underset{\sim}{\zeta}}, \underset{\sim}{\zeta})$. The coordinate process Z, together with a σ-finite measure Q constructed by "knitting" together the transition functions for X and $\hat{X}$ via the duality measure ξ, constitute the auxiliary process. X and $\hat{X}$ embed naturally into this process, and the switching identity (0.2) expresses itself in the *shift invariance* of Q:

$$Q(Y) = Q(Y \circ \underset{\sim}{\theta}_t), \quad t \in \mathbb{R}, \quad Y \in \underset{\sim}{F}^o(-\infty,\infty). \tag{0.3}$$

Here $\underset{\sim}{F}^o(-\infty,\infty) = \sigma(Z_t : -\infty < t < \infty)$, and $\underset{\sim}{\theta}_t$ is the shift operator on $\underset{\sim}{\Omega}$. This shows that $\{\underset{\sim}{\theta}_t\}$ is a *flow* on $(\underset{\sim}{\Omega}, \underset{\sim}{F}^o(-\infty,\infty), Q)$: a one-parameter group (under composition) of bimeasurable bijections $\underset{\sim}{\theta}_t : \underset{\sim}{\Omega} \to \underset{\sim}{\Omega}$ with identity $\underset{\sim}{\theta}_0$.

We reproduce here a few definitions and facts from [12] and [13]; the reader should consult those sources for a fuller exposition. On $\underset{\sim}{\Omega}$ the σ-algebra $\sigma(Z_u : s \le u \le t)$ is denoted $\underset{\sim}{F}^o[s,t]$; when completed by adjoining all Q-null sets, the 0-superscript is deleted. The resulting families $(\underset{\sim}{F}(-\infty,t])$ and $(\underset{\sim}{F}[s,\infty))$ are right and left continuous respectively; [11]. For each $x \in E$ we have measures $\underset{\sim}{P}^x$ on $\underset{\sim}{F}[0,\infty)$ and $\hat{\underset{\sim}{P}}^x$ on $\underset{\sim}{F}(-\infty,0]$: the measure $\underset{\sim}{P}^x$ is the image of P^x under the natural projection of $s \to Z_s$ $(s \geq 0,\ Z_0 \in E)$ onto $s \to X_s$ and is carried by $\{\underset{\sim}{\omega}: Z_0(\underset{\sim}{\omega}) = x\}$; $\hat{\underset{\sim}{P}}^x$ is defined analogously from $\hat{P}^x$ and is carried by $\{\underset{\sim}{\omega}: Z_{0-}(\omega) = x\}$. These measures appear in formulations of the Markov properties of Z and in the following explicit expressions for the optional and cooptional projections Π^+W and Π^-W of a bounded $\mathcal{B}(\mathbb{R}) \otimes \underset{\sim}{F}^{o}(-\infty,\infty)$-measurable process W (valid for $\hat{\zeta}(\underset{\sim}{\omega}) < t < \zeta(\underset{\sim}{\omega})$):

$$\text{(0.4)}\quad \text{(i)}\qquad \Pi^+W(t,\underset{\sim}{\omega}) = \int W_t(\underset{\sim}{\omega}|t|\underset{\sim}{\theta}_{-t}\underset{\sim}{\omega}')\ \underset{\sim}{P}^{Z_t(\underset{\sim}{\omega})}(d\underset{\sim}{\omega}')$$

$$\text{(ii)}\qquad \Pi^-W(t,\underset{\sim}{\omega}) = \int W_t(\underset{\sim}{\theta}_{-t}\underset{\sim}{\omega}'|t|\underset{\sim}{\omega})\ \hat{\underset{\sim}{P}}^{Z_{t-}(\underset{\sim}{\omega})}(d\underset{\sim}{\omega}').$$

(The "splicing map" $(\cdot|s|\cdot): \underset{\sim}{\Omega} \to \underset{\sim}{\Omega}$ is given by

$$Z_t(\underset{\sim}{\omega}|s|\underset{\sim}{\omega}') = Z_t(\underset{\sim}{\omega}) \quad \text{if} \quad t \vee \hat{\zeta}(\underset{\sim}{\omega}') < s \leq \zeta(\underset{\sim}{\omega}) \quad \text{or} \quad s > \zeta(\underset{\sim}{\omega})$$

$$= Z_t(\underset{\sim}{\omega}') \text{ if } s \leq \zeta(\underset{\sim}{\omega}) \text{ and either } s \leq t \text{ or } s < \hat{\zeta}(\underset{\sim}{\omega}').)$$

Formulas (0.4i,ii) were derived in [11], where optional and cooptional were called "right" and "left." In the next section these formulas form the basis of some calculations which lead to representation results for processes and purely atomic homogeneous random measures which are both optional and cooptional.

1. Representation Results for Optional and Cooptional Processes and Random Measures

Using (0.4) an easy computation proves that Π^+ and Π^- commute: if $W \in \mathcal{B}(\mathbb{R}) \otimes \underset{\sim}{F}^{o}(-\infty,\infty)$, Π^+W and Π^-W are again $\mathcal{B}(\mathbb{R}) \otimes \underset{\sim}{F}^{o}(-\infty,\infty)$ -

measurable, and for $\hat{\zeta} < t < \zeta$,

$$\Pi^- \Pi^+ W(t,\omega) = \int \Pi^+ W_t(\theta_{-t}\omega' | t | \omega) \, \hat{P}^{Z_{t-}(\omega)}(d\omega')$$

$$= \int [\int W_t([\theta_{-t}\omega' | t | \omega] | t | \theta_{-t}\omega'') P^{Z_t(\theta_{-t}\omega' | t | \omega)}(d\omega'') \hat{P}^{Z_{t-}(\omega)}(d\omega')$$

$$= \int\int W_t \circ \theta_{-t}(\omega' | 0 | \omega'') P^{Z_t(\omega)}(d\omega'') \hat{P}^{Z_{t-}(\omega)}(d\omega') = \Pi^+ \Pi^- W(t,\omega).$$

Motivated by this calculation, define a measure ${}^xQ^y$ on $(\Omega, F^o(-\infty,\infty))$ by

(1.1) $$ {}^xQ^y(Y) = \int\int Y(\omega' | 0 | \omega'') \hat{P}^x(d\omega') P^y(d\omega'').$$

It is easy to see that ${}^xQ^y$ is a measure on the trace of $F^o(-\infty,\infty)$ on $\{Z_{0-} \in E, Z_0 \in E\}$ which is carried by $\{Z_{0-} = x, Z_0 = y\}$ and that $(x,y) \to {}^xQ^y(Y)$ is in $E \otimes E$ for $Y \in F^o(-\infty,\infty)$. Let Π stand for the composition of the optional and cooptional projections. Then, for $\hat{\zeta} < t < \zeta$,

(1.2) $$\Pi W(t,\omega) = {}^{Z_{t-}(\omega)}Q^{Z_t(\omega)}(W_t \circ \theta_{-t}).$$

This formula shows that ΠW is indistinguishable from a process of the form $t \to f(t, Z_{t-}, Z_t)$ on $(\hat{\zeta}, \zeta)$ where $f \in B(\mathbb{R}) \otimes E \otimes E$: take $f(t,x,y) = \int W(t, \theta_{-t}\omega) \, {}^xQ^y(d\omega)$. If W is homogeneous, ΠW is indistinguishable from $f(Z_{t-}, Z_t)$ where $f(x,y) = {}^xQ^y(W_0) \in E \otimes E$.

Now suppose λ is a finite random measure which is both optional and cooptional (in the sense of [11]). In [11] it is shown that, if γ is any random measure carried by $(\hat{\zeta}, \zeta)$ for which the dual optional projection $\hat{\Pi}^+\gamma$ is defined, then the processes $t \to \hat{\Pi}^+\gamma\{t\}$ and $t \to \Pi^+(\gamma\{t\})$ are indistinguishable (the latter is the optional projection of the process $(t,\omega) \to \gamma(\omega,\{t\})$). Of course the same statement with cooptional projections replacing optional ones holds equally well. For λ this implies that the process

$$J(t,\underset{\sim}{\omega}) = \lambda(\underset{\sim}{\omega},\{t\}) \tag{1.3}$$

is both optional and cooptional, and homogeneous if λ is. We may conclude that $\lambda(\underset{\sim}{\omega},\{t\})$ is indistinguishable from $f(t,Z_{t-},Z_t)$ as above.

In the case of purely atomic homogeneous random measures, this result contains the representations for purely discontinuous A.F.'s found by Revuz and Sharpe under strong duality hypotheses for natural and "quasi-left-continuous" A.F.'s respectively. If λ is "natural," i.e., $\lambda\{t\} = 0$ if $Z_{t-} \neq Z_t$, the corresponding f is necessarily carried by the diagonal in $E \times E$, and $\{x: f(x,x) > 0\}$ is semipolar (visited only countably often by the process Z) [1]. If λ is quasi-left-continuous, i.e., $\lambda\{t\} = 0$ if $Z_{t-} = Z_t$, then f vanishes on the diagonal. These results pass over to results for X and $\hat{X}$ via the embedding techniques of [12].

REMARK. The natural way to work with "additive functionals" when using the auxiliary process is to consider homogeneous random measures: an additive functional A of X (or $\hat{X}$) corresponds to the random measure λ for which A is the "distribution function," and λ in turn embeds naturally into a homogeneous random measure $\underset{\sim}{\lambda}$ over Z (see [12]). Results of Meyer [10] allow us to assume that our additive functionals are perfect and satisfy $A_t = A_t \circ k_u$ if $t < u$ (here k_u is the "killing operator"). These assumptions guarantee that the corresponding random measure $\underset{\sim}{\lambda}$ over Z will be both optional and cooptional. In addition to the property mentioned above, an optional and cooptional random measure γ will satisfy the following: the mapping

$$\underset{\sim}{\omega} \to \gamma(\underset{\sim}{\omega},[a,b])$$

is in $\underset{\sim}{F}[a,b]$, and

$$Q \int W_t \, \gamma(dt) = Q \int \Pi W_t \, \gamma(dt)$$

for all $W \in b(\mathcal{B}(\mathbb{R}) \otimes \underset{\sim}{F}(-\infty,\infty))$.

2. A General Switching Identity

In this section we prove a general switching identity for random measures that are both optional and cooptional and have σ-finite spectral measures (see below for terminology). Various formulas are mere consequences of this identity. From one of these formulas we see a natural way to define the α-potential of a measure in the case of weak duality (i.e., in the absence of an α-potential density $u^\alpha(x,y)$). A switching identity for potentials of measures then follows. Another formula makes explicit the connection between the spectral measure and the Revuz measure for certain homogeneous random measures which correspond to additive functionals for X and $\hat{X}$. (Section 3 shall be devoted to the verification of the hypotheses of this section for certain homogeneous random measures.)

(2.1) ASSUMPTIONS. All random measures λ shall be subject to the following assumptions in this section:

(A1) λ is optional and cooptional.

(A2) λ is carried by $(\hat{\underset{\sim}{\zeta}},\underset{\sim}{\zeta})$.

(A3) There exists a σ-finite measure $\bar{\mu}_\lambda(dt,dx,dy)$, called the *spectral measure* of λ (following Dynkin) such that, for every $f \in (\mathcal{B}(\mathbb{R}) \otimes \mathcal{E} \otimes \mathcal{E})^+$,

$$Q \int f(t,Z_{t-},Z_t) \, \lambda(dt) = \int f(t,x,y) \, \bar{\mu}_\lambda(dt,dx,dy).$$

Since all processes that are both optional and cooptional are indistinguishable from $f(t,Z_{t-},Z_t)$ for $\hat{\underset{\sim}{\zeta}} < t < \underset{\sim}{\zeta}$, it follows (just as in [4]) that random measures satisfying (A1)-(A3) are determined by their spectral measures.

(2.2) DEFINITIONS. Suppose λ satisfies (A1)-(A3) and $f \in (\mathcal{B}(\mathbb{R}) \otimes \mathcal{E} \otimes \mathcal{E})^+$.

$$U_\lambda f(z) \equiv \underset{\sim}{E}^z \int_0^\infty f(t,Z_{t-},Z_t)\,\lambda(dt)$$

$$f\hat{U}_\lambda(z) \equiv \hat{\underset{\sim}{E}}^z \int_{-\infty}^0 f(t,Z_{t-},Z_t)\,\lambda(dt).$$

When $f \equiv 1$, we write u_λ for $U_\lambda 1$ and $\hat{u}_\lambda$ for $1\hat{U}_\lambda$.

(2.3) NOTATION. For $-\infty \le a \le b \le \infty$, $\lambda(a,b)$ shall stand for $\lambda((a,b))$. Also we adopt the convention that, for any measure $\ell(dt)$ on $\mathbb{R}$, $\int_a^b f(t)\,\ell(dt)$ shall stand for $\int 1_{(a,b)}f(t)\,\ell(dt)$.

(2.4) THEOREM (General switching identity). *Suppose* γ *and* λ *satisfy* (A1)-(A3). *Then,*

$$Q(\gamma(-\infty,0)\,\lambda(0,\infty)\,1_{\{Z_0 \in E\}})$$

$$= \int 1_{(-\infty,0)}(t)\,P_{-t}u_\lambda(y)\,\bar{\mu}_\gamma(dt,dx,dy)$$

$$= \int 1_{(0,\infty)}(t)\,\hat{u}_\gamma\hat{P}_t(x)\,\bar{\mu}_\lambda(dt,dx,dy).$$

PROOF. We first compute optional and cooptional projections of certain processes. Let

$$Y_t = 1_{(-\infty,0)}(t)\,1_{(\hat{\underset{\sim}{\zeta}},\underset{\sim}{\zeta})}(t)\,\lambda(0,\infty)\,1_{\{Z_0 \in E\}}.$$

Since λ is cooptional, $\lambda(0,\infty) \in \underset{\sim}{F}(0,\infty)$. It follows from (0.4) (see also [4] Lemma 3.2) that

$$\Pi^+ Y_t = 1_{(-\infty,0)}(t)\ P_{-t} u_\lambda(Z_t).$$

Next let $W_t = 1_{(0,\infty)}(t)\ 1_{(\underset{\sim}{\hat{\zeta}},\underset{\sim}{\zeta})}(t)\ \gamma(-\infty,0)\ 1_{\{Z_0 \in E\}}$. Since γ is optional it follows that $\gamma(-\infty,0)$ is in $\underset{\sim}{F}(-\infty,0)$. Again by (0.4), we have $\Pi^- W_t = 1_{(0,\infty)}(t)\ \hat{u}_\gamma \hat{P}_t(Z_{t-})$.

Using the fact that γ is optional again, we have

$$\begin{aligned} Q[\gamma(-\infty,0)\ \lambda(0,\infty)] &= Q \int (1_{(-\infty,0)}(t)\ 1_{(\underset{\sim}{\hat{\zeta}},\underset{\sim}{\zeta})}(t)\ \lambda(0,\infty))\ \gamma(dt) \\ &= Q \int_{-\infty}^{0} P_{-t} u_\lambda(Z_t)\ \gamma(dt) \\ &= \int 1_{(-\infty,0)}(t)\ P_{-t} u_\lambda(y)\ \bar{\mu}_\gamma(dt,dx,dy). \end{aligned}$$

Using the fact that λ is cooptional we have,

$$\begin{aligned} Q[\gamma(-\infty,0)\ \lambda(0,\infty)] &= Q \int (1_{(0,\infty)}(t)\ 1_{(\underset{\sim}{\hat{\zeta}},\underset{\sim}{\zeta})}(t)\ \gamma(-\infty,0))\ \lambda(dt) \\ &= Q \int_{0}^{\infty} \hat{u}_\gamma \hat{P}_t(Z_{t-})\ \lambda(dt) \\ &= \int 1_{(0,\infty)}(t)\ \hat{u}_\gamma \hat{P}_t(x)\ \bar{\mu}_\lambda(dt,dx,dy). \end{aligned}$$

(2.5) COROLLARY. *Suppose* γ *and* λ *satisfy* (A1)-(A3) *and* $f,g \in (\mathcal{B}(\mathbb{R}) \otimes \mathcal{E} \otimes \mathcal{E})^+$. *Then*

$$\begin{aligned} &\int 1_{(-\infty,0)}(t)\ f(t,x,y)\ P_{-t} U_\lambda g(y)\ \bar{\mu}_\gamma(dt,dx,dy) \\ &\qquad = \int 1_{(0,\infty)}(t)\ g(t,x,y)\ f\hat{U}_\gamma \hat{P}_t(x)\ \bar{\mu}_\lambda(dt,dx,dy). \end{aligned}$$

PROOF. We may assume f, g bounded. Then, (2.5) follows by applying (2.4) to the random measures $f(t,Z_{t-},Z_t)\ \gamma(dt)$ and $g(t,Z_{t-},Z_t)\ \lambda(dt)$ with spectral measures $f(t,x,y)\ \bar{\mu}_\gamma(dt,dx,dy)$ and $g(t,x,y)\ \bar{\mu}_\lambda(dt,dx,dy)$ respectively. This completes the proof.

We will now consider the following assumptions of homogeneity.

(2.6) ASSUMPTIONS OF HOMOGENEITY.

(H1) For every $t \in \mathbb{R}$, $\underset{\sim}{\omega}$, and real Borel set I,

$$\lambda(\underset{\sim}{\theta}_t\underset{\sim}{\omega}, I) = \lambda(\underset{\sim}{\omega}, I+t).$$

(H2) There is a σ-finite measure $\nu_\lambda(dx,dy)$ such that

$$\bar{\mu}_\lambda\ (dt,dx,dy) = dt\ \nu_\lambda(dx,dy).$$

(2.7) DEFINITION. Suppose λ satisfies (A1)-(A3), $f \in (E \otimes E)^+$, and $a \geq 0$.

$$U^\alpha_\lambda f(z) \equiv \underset{\sim}{E}^z \int_0^\infty e^{-\alpha t}\ f(Z_{t-},Z_t)\ \lambda(dt)$$

$$f\hat{U}^\alpha_\lambda(z) \equiv \hat{\underset{\sim}{E}}^z \int_{-\infty}^0 e^{\alpha t}\ f(Z_{t-},Z_t)\ \lambda(dt).$$

When $f \equiv 1$ we write u^α_λ for $U^\alpha_\lambda 1$ and $\hat{u}^\alpha_\lambda$ for $1\hat{U}^\alpha_\lambda$.

(2.8) COROLLARY. *Suppose* γ *and* λ *satisfy* (A1)-(A3), (H1)-(H2), *and* $\alpha,\beta \geq 0$. *Then,*

$$\int U^\beta u^\alpha_\lambda(y)\ \nu_\gamma(dx,dy) = \int \hat{u}^\beta_\gamma \hat{U}^\alpha(x)\ \nu_\lambda(dx,dy).$$

PROOF. Let $f(t,x,y) = e^{\beta t}$, $g(t,x,y) = e^{-\alpha t}$. Then $U_\lambda g = u^\alpha_\lambda$ and $f\hat{U}_\gamma = \hat{u}^\beta_\gamma$. Applying (2.5) to this f and g we have

$$\int_{E\times E} \int_{-\infty}^{0} e^{\beta t} P_{-t} u^\alpha_\lambda(y)dt \ \nu_\gamma(dx,dy) = \int_{E\times E} \int_{0}^{\infty} e^{-\alpha t} \hat{u}^\beta_\gamma \hat{P}_t(x)dt \ \nu_\lambda(dx,dy),$$

and this implies our result.

(2.9) THEOREM. *Suppose* γ *and* λ *satisfy* (A1)-(A3), (H1)-(H2), *and* $\alpha \geq 0$. *Then*

$$\int u^\alpha_\lambda(y) \ \nu_\gamma(dx,dy) = \int \hat{u}^\alpha_\gamma(x) \ \nu_\lambda(dx,dy).$$

PROOF. Fix x and $\beta > 0$. Then

$$\begin{aligned}
\beta\hat{u}^\alpha_\gamma \hat{U}^{\alpha+\beta}(x) &= \int_{-\infty}^{0} \beta e^{(\alpha+\beta)t} \ \hat{u}^\alpha_\gamma \hat{P}_{-t}(x)dt \\
&= \int_{\infty}^{0} \beta e^{(\alpha+\beta)t} \ \hat{\underset{\sim}{E}}^x[\hat{\underset{\sim}{E}}^{Z_t-} \int_{-\infty}^{0} e^{\alpha s} \ \gamma(ds)]dt \\
&= \int_{-\infty}^{0} \beta e^{(\alpha+\beta)t} \ \hat{\underset{\sim}{E}}^x[(\int_{-\infty}^{0} e^{\alpha s} \ \gamma(ds)) \circ \underset{\sim}{\theta}_t]dt \\
&= \int_{-\infty}^{0} \beta e^{(\alpha+\beta)t} \ \hat{\underset{\sim}{E}}^x[\int_{-\infty}^{t} e^{\alpha(s-t)} \ \gamma(ds)]dt \\
&= \int_{-\infty}^{0} e^{\beta t} \ \hat{\underset{\sim}{E}}^x[\int_{-\infty}^{t} e^{\alpha s} \ \gamma(ds)]dt \\
&= \int_{-\infty}^{0} e^{t} \ \hat{\underset{\sim}{E}}^x[\int_{-\infty}^{t/\beta} e^{\alpha s} \ \gamma(ds)]dt.
\end{aligned}$$

Thus $\beta\hat{u}^\alpha_\gamma \hat{U}^{\alpha+\beta}(x) \uparrow \hat{u}^\alpha_\gamma(x)$ as $\beta \uparrow \infty$.

Next, fix y, $\beta > 0$. Then

$$\beta U^\alpha u^{\alpha+\beta}_\lambda(y) = \int_0^\infty \beta e^{-\alpha t} \ \underset{\sim}{E}^y[\underset{\sim}{E}^{Z_t} \int_0^\infty e^{-(\alpha+\beta)s} \ \lambda(ds)]dt$$

$$= \int_0^\infty \beta e^{-\alpha t} \underset{\sim}{E}^y[(\int_0^\infty e^{-(\alpha+\beta)s} \lambda(ds)) \circ \underset{\sim}{\theta}_t]dt$$

$$= \int_0^\infty \beta e^{-\alpha t} \underset{\sim}{E}^y[\int_t^\infty e^{-(\alpha+\beta)(s-t)} \lambda(ds)]dt$$

$$= \int_0^\infty \beta e^{\beta t} \underset{\sim}{E}^y[\int_t^\infty e^{-(\alpha+\beta)s} \lambda(ds)]dt$$

$$= \underset{\sim}{E}^y \int_0^\infty \beta e^{\beta t}\Big(\int_t^\infty e^{-(\alpha+\beta)s} \lambda(ds)\Big) dt.$$

Thus,

$$\int \beta U^\alpha u_\lambda^{\alpha+\beta}(y)\, \nu_\gamma(dx,dy) = Q \int_0^1 \beta U^\alpha u_\lambda^{\alpha+\beta}(Z_u)\, \gamma(du)$$

$$= Q \int_0^1 \Big(\int_0^\infty \beta e^{\beta t}\big(\int_t^\infty e^{-(\alpha+\beta)s} \lambda(ds)\big)dt\Big) \circ \underset{\sim}{\theta}_u\, \gamma(du)$$

$$= Q \int_0^1 \Big(\int_0^\infty \beta e^{\beta t}\big(\int_{t+u}^\infty e^{-(\alpha+\beta)(s-u)} \lambda ds)\big)dt\Big)\, \gamma(du)$$

$$= Q \int_0^1 \Big(\int_u^\infty e^{-(\alpha+\beta)(s-u)}\big(\int_0^{s-u} \beta e^{\beta t} dt\big)\, \lambda(ds)\Big)\gamma(du)$$

$$= Q \int_0^1 \Big(\int_u^\infty (e^{-\alpha(s-u)} - e^{-(\alpha+\beta)(s-u)})\lambda(ds)\Big)\, \gamma(du)$$

$$= Q \int_0^1 \Big(\int_0^\infty (e^{-\alpha s} - e^{-(\alpha+\beta)s})\, \lambda(ds)\Big) \circ \underset{\sim}{\theta}_u\, \gamma(du)$$

$$= Q \int_0^1 \underset{\sim}{E}^{Z_u}\Big(\int_0^\infty (e^{-\alpha s} - e^{-(\alpha+\beta)s})\, \lambda(ds)\Big)\, \gamma(du).$$

Note that in the above calculation we used the fact that Q - a.s. λ is σ-finite in order to interchange the order of integration.

Now, by (2.8) we have

$$\int \beta U^\alpha u_\lambda^{\alpha+\beta}(y)\, \nu_\gamma(dx,dy) = \int \beta \hat{u}_\gamma^\alpha \hat{U}^{\alpha+\beta}(x)\, \nu_\lambda(dx,dy).$$

Using the above two calculations and then letting $\beta \uparrow \infty$ we obtain the result.

(2.10) COROLLARY. *Suppose* γ *and* λ *satisfy* (A1)-(A3), (H1)-(H2), *and* $\alpha \geq 0$, *and* $f,g \in (E \otimes E)^+$. *Then,*

$$\int U_\lambda^\alpha g(y)\ f(x,y)\ \nu_\gamma(dx,dy) = \int f\hat{U}_\gamma^\alpha(x)\ g(x,y)\ \nu_\lambda(dx,dy).$$

PROOF. We can assume f, g bounded. Then $f(Z_{t-},Z_t)\ \gamma(dt)$ and $g(Z_{t-},Z_t)\ \lambda(dt)$ satisfy (A1)-(A3), (H1)-(H2) and have spectral measures $dt\ f(x,y)\ \nu_\gamma(dx,dy)$ and $dt\ g(x,y)\ \nu_\lambda(dx,dy)$ respectively. Applying (2.9) we are done.

(2.11) COROLLARY. *Suppose* λ *satisfies* (A1)-(A3), (H1)-(H2), $f \in (E \otimes E)^+$, $g \in E^+$. *Then,*

$$\int g(y)\ U_\lambda^\alpha f(y)\ \xi(dy) = \int g\hat{U}^\alpha(x)\ f(x,y)\ \nu_\lambda(dx,dy),$$

and

$$\int g(x)\ f\hat{U}_\lambda^\alpha(x)\ \xi(dx) = \int U^\alpha g(y)\ f(x,y)\ \nu_\lambda(dx,dy).$$

PROOF. Let $\gamma(dt) = 1_{(\hat{\underset{\sim}{\zeta}},\underset{\sim}{\zeta})}(t)\ dt$ and $\ell \in (\mathcal{B}(\mathbb{R}) \otimes E \otimes E)^+$. Then,

$$\begin{aligned} Q \int \ell(t,Z_{t-},Z_t)\ \gamma(dt) &= Q \int_{-\infty}^{\infty} \ell(t,Z_{t-},Z_t)\ 1_{(\hat{\underset{\sim}{\zeta}},\underset{\sim}{\zeta})}(t)dt \\ &= \int_{-\infty}^{\infty} Q[\ell(t,Z_{t-},Z_t)\ 1_{(\hat{\underset{\sim}{\zeta}},\underset{\sim}{\zeta})}(t)]dt \\ &= \int_{-\infty}^{\infty} \int_E \ell(t,x,x)\ \xi(dx)dt. \end{aligned}$$

That is to say that, if $h \in (E \otimes E)^+$, then $\int h(x,y)\ \nu_\gamma(dx,dy) = \int h(x,x)\ \xi(dx)$.

Now, if we set $k(x,y) = g(y)$, then we have

$$k\hat{U}^{\alpha}_{\gamma}(x) = \hat{\underset{\sim}{E}}^{x} \int_{-\infty}^{0} e^{\alpha t}\, g(Z_t)dt = \int_{-\infty}^{0} e^{\alpha t}\, \hat{P}_{-t}\, g(x)dt = g\hat{U}^{\alpha}(x).$$

By (2.10) we have

$$\int U^{\alpha}_{\lambda}f(y)\, k(x,y)\, \nu_{\gamma}(dx,dy) = \int k\hat{U}^{\alpha}_{\gamma}(x)\, f(x,y)\, \nu_{\lambda}(dx,dy),$$

and thus the first formula of our statement is proved. The second follows similarly.

Corollary (2.11) provides a way of computing $\nu_{\lambda}(dx,dy)$ intrinsically, i.e. using only the original processes X and $\hat{X}$.

(2.12) THEOREM. *Suppose* λ *satisfies* (A1)-(A3) *and* (H1)-(H2), *and* $f \in (E \otimes E)^{+}$. *Then,*

$$\int f(x,y)\, \nu_{\lambda}(dx,dy) = \lim_{\alpha\to\infty} \alpha \int U^{\alpha}_{\lambda}f(y)\, \xi(dy)$$

$$= \lim_{\alpha\to\infty} \alpha \int f\hat{U}^{\alpha}_{\lambda}(x)\, \xi(dx).$$

Both of these limits are increasing with α.

PROOF. This follows directly from the fact that $\alpha U^{\alpha}1, \alpha 1\hat{U}^{\alpha} \uparrow 1$ as $\alpha \uparrow \infty$, combined with (2.11) letting $g = 1$.

Theorem (2.12) is the connection with the theory of Revuz measures mentioned at the beginning of this section.

We now digress briefly to compare some of these results with well known facts in the case where the semi-groups P_t and $\hat{P}_t$ have a jointly measurable density $p_t(x,y)$ relative to ξ. In particular we mean that for every $f \in E^+$,

$$P_t f(x) = \int f(y)\, p_t(x,y)\, \xi(dy), \quad f\hat{P}_t(y) = \int f(x)\, p_t(x,y)\, \xi(dx).$$

We also mean explicitly that p_t satisfies the Chapman-Kolmogorov equation

$$p_{t+s}(x,y) = \int p_s(x,z)\, p_t(z,y)\, \xi(dz).$$

The following uses an argument in [4] (see the proof of Theorem 4.1 there) adapted to our situation.

(2.13) THEOREM. *Assume* $P_t, \hat{P}_t$ *have a density* p_t *relative to* ξ *in the manner described above. Next, suppose* λ *satisfies* (A1)-(A3). *Then for each* x, (see (2.2) also)

$$u_\lambda(x) = \int 1_{(0,\infty)}(t)\, p_t(x,y)\, \bar{\mu}_\lambda(dt,dy,dz),$$

$$\hat{u}_\lambda(x) = \int 1_{(-\infty,0)}(t) p_{-t}(y,x)\, \bar{\mu}_\lambda(dt,dz,dy).$$

PROOF. Fix $u > 0$, $u < t_1 < t_2 < \cdots < t_n$, $f_1, \ldots, f_n \in bE^+$. Then

$$\underset{\sim}{E}^x f_1(Z_{t_1}) \cdots f_n(Z_{t_n}) = \underset{\sim}{E}^x h(Z_{t_1}),$$

where

$$h(y) = \underset{\sim}{E}^y f_1(Z_0)\, f_2(Z_{t_2-t_1}) \cdots f_n(Z_{t_n-t_1}).$$

Now

$$\underset{\sim}{E}^x h(Z_{t_1}) = P_{t_1} h(x) = P_u P_{t_1-u} h(x)$$

$$= \int p_u(x,z)\, P_{t_1-u}h(z)\, \xi(dz) = Q[p_u(x,Z_u)\, P_{t_1-u}h(Z_u)]$$

$$= Q[p_u(x,Z_u)\, h(Z_{t_1})] = Q[p_u(x,Z_u)\, f_1(Z_{t_1})\cdots f_n(Z_{t_n})].$$

(The last two equalities are consequences of the Markov property of the measure Q.) By a monotone class argument it follows that $\underset{\sim}{E}^x\lambda(u,\infty)\, 1_{\{u<\underset{\sim}{\zeta}\}} = Q[p_u(x,Z_u)\, \lambda(u,\infty)]$ since $\lambda(u,\infty) \in \underset{\sim}{F}(u,\infty)$. Thus

$$\underset{\sim}{E}^x\lambda(u,\infty)\, 1_{\{u<\underset{\sim}{\zeta}\}} = Q[p_u(x,Z_{u-})\, \lambda(u,\infty)]$$

$$= Q \int (1_{(u,\infty)}(t)\, p_u(x,Z_{u-}))\, \lambda(dt)$$

$$= Q \int \Pi^-(1_{(u,\infty)}(t)\, p_u(x,Z_{u-})\, 1_{(\hat{\underset{\sim}{\zeta}}<t<\underset{\sim}{\zeta})})\, \lambda(dt)$$

$$= Q \int 1_{(u,\infty)}(t)\, (p_u(x,\cdot)\, \hat{P}_{t-u})(Z_{t-})\, \lambda(dt)$$

$$= \int 1_{(u,\infty)}(t)\, (p_u(x,\cdot)\, \hat{P}_{t-u})(y)\, \bar{\mu}_\lambda(dt,dy,dz).$$

But

$$p_u(x,\cdot)\, \hat{P}_{t-u}(y) = \int p_u(x,z)\, p_{t-u}(z,y)\, \xi(dz) = p_t(x,y).$$

Thus we have

$$\underset{\sim}{E}^x\lambda(u,\infty)\, 1_{\{u<\underset{\sim}{\zeta}\}} = \int 1_{(u,\infty)}(t)\, p_t(x,y)\, \bar{\mu}_\lambda(dt,dy,dz).$$

Let $u \downarrow 0$ to get the desired first formula. The second formula follows in a similar manner.

(2.14) COROLLARY. *We make the same assumption as in* (2.13) *and let* $f \in (\mathcal{B}(\mathbb{R}) \otimes \mathcal{E} \otimes \mathcal{E})^+$. *For each* x,

$$U_\lambda f(x) = \int 1_{(0,\infty)}(t)\, p_t(x,y)\, f(t,y,z)\, \bar{\mu}_\lambda(dt,dy,dz),$$

$$f\hat{U}_\lambda(x) = \int 1_{(-\infty,0)}(t)\, p_{-t}(y,x)\, f(t,z,y)\, \bar{\mu}_\lambda(dt,dz,dy).$$

PROOF. This follows directly from (2.13) in the same manner, for example, that (2.5) follows from (2.4).

(2.15) COROLLARY. *We again make the same assumptions as in* (2.13) *with the addition now of* (H1)-(H2). *Let* $\alpha \geq 0$. *Then,*

$$u_\lambda^\alpha = U^\alpha\, \nu_\lambda^1, \quad \hat{u}_\lambda^\alpha = \nu_\lambda^2\, \hat{U}^\alpha,$$

where, for every h,

$$\int h(y)\, \nu_\lambda^1(dy) \equiv \int h(y)\, \nu_\lambda(dy,dz),$$

$$\int h(z)\, \nu_\lambda^2(dz) \equiv \int h(z)\, \nu_\lambda(dy,dz).$$

PROOF. In (2.14) let $f(t,y,z) = e^{-\alpha t}$ in the first formula and $f(t,z,y) = e^{\alpha t}$ in the second.

Theorem (2.9) now can be thought of as an extension of the straightforward switching identity which can be proved by Fubini's theorem in the case of strong duality, namely,

$$\int U^\alpha\, \nu_\lambda^1(y)\, \nu_\gamma(dx,dy) = \int \nu_\gamma^2 \hat{U}^\alpha(x)\, \nu_\lambda(dx,dy).$$

Written in a more compact form (with obvious notation):

$$\langle \nu_\lambda^2, U^\alpha \nu_\lambda^1 \rangle = \langle \nu_\gamma^2 \hat{U}^\alpha, \nu_\lambda^1 \rangle .$$

We now return to the general discussion.

(2.16) DEFINITION. Suppose λ satisfies (A1)-(A3), (H1)-(H2). We say that λ is *natural*, following Dynkin [4], if there exists a σ-finite measure, which we also will denote by ν_λ, on $(E,\mathcal{E})$ such that for every $h \in (\mathcal{E} \otimes \mathcal{E})^+$,

$$\int h(x,y) \quad \nu_\lambda(dx,dy) = \int h(y,y) \quad \nu_\lambda(dy).$$

Of course, this is simply the condition that $\nu_\lambda(dx,dy)$ be carried by the diagonal in $E \times E$.

In the case of densities our previous discussion leads to an interpretation of (2.9), in the case where λ and γ are natural, as a switching identity for potentials of measures:

$$\langle \nu_\gamma, U^\alpha \nu_\lambda \rangle = \langle \nu_\gamma \hat{U}^\alpha, \nu_\lambda \rangle .$$

This makes it "natural" to define, in the case of weak duality, $U^\alpha \nu_\lambda \equiv u^\alpha_\lambda$, where λ is natural, and $\nu_\lambda \hat{U}^\alpha \equiv \hat{u}^\alpha_\lambda$.

Now suppose in (2.10) that γ and λ are natural. In this case, for $f \in \mathcal{E}^+$ we define (with a slight abuse of notation)

$$U^\alpha_\lambda f(z) = \underset{\sim}{E}^z \int_0^\infty e^{-\alpha t} f(Z_t) \lambda(dt),$$

$$f\hat{U}^\alpha_\lambda(Z) = \hat{\underset{\sim}{E}}^z \int_{-\infty}^0 e^{\alpha t} f(Z_{t-}) \lambda(dt).$$

Corollary (2.10) now becomes: For $f,g \in \mathcal{E}^+$, $\alpha \geq 0$,

$$\int U^\alpha_\lambda g(y) f(y) \nu_\gamma(dy) = \int g(y) f\hat{U}^\alpha_\gamma(y) \nu_\lambda(dy).$$

In the case where $\gamma = \lambda$ this becomes

$$\langle f, U^\alpha_\lambda g\rangle = \langle f\hat{U}^\alpha_\lambda, g\rangle ,$$

where $\langle h, k\rangle$ denotes $\int hk\, d\nu_\lambda$. Thus U^α_λ and $\hat{U}^\alpha_\lambda$ are in weak duality with respect to ν_λ.

3. Measures Associated to Homogeneous Random Measures

As pointed out at the end of section 1, it is *homogeneous* random measures which correspond to additive functionals; from now on we restrict our attention to this class. Most of this section will be devoted to showing that *finite* random measures satisfying (H1) will also satisfy (H2). A random measure λ is called *finite* provided

$$Q\{\underset{\sim}{\omega}:\ \lambda(\underset{\sim}{\omega}, I) = \infty;\ \ I \subset (\hat{\underset{\sim}{\zeta}}(\underset{\sim}{\omega}),\ \underset{\sim}{\zeta}(\underset{\sim}{\omega}))\} = 0$$

for any closed interval I. We begin by introducing another useful measure associated to a homogeneous random measure. Actually there are three measures which are useful to consider in connection with a homogeneous random measure λ over Z. The first, familiar to cognoscenti of the general theory of processes for its role in defining the dual projection of a random measure, is the basic measure μ_λ on $\mathcal{B}(\mathbb{R}) \otimes \underset{\sim}{F}(-\infty,\infty)$ given by

$$\mu_\lambda(W) = Q \int W_t\, \lambda(dt) \quad \text{for} \quad W \in \mathcal{B}(\mathbb{R}) \otimes \underset{\sim}{F}(-\infty,\infty). \tag{3.1}$$

The second will be of interest in what follows. It is exactly the "Palm measure" associated to λ in the theory of flows: the measure $\underset{\sim}{\mu}_\lambda$ on $(\underset{\sim}{\Omega},\ \underset{\sim}{F}(-\infty,\infty))$ given by

(3.2) $$\underset{\sim}{\mu}_\lambda(A) = Q \int_0^1 1_A \circ \underset{\sim}{\theta}_s \, \lambda(ds).$$

When integrability conditions on λ guarantee that $\underset{\sim}{\mu}_\lambda$ is σ-finite, the homogeneity of λ together with the shift invariance (0.3) of Q yield

(3.3) $$\int Q(d\underset{\sim}{\omega}) \int f(\underset{\sim}{\theta}_t \underset{\sim}{\omega}, t)\, \lambda(\underset{\sim}{\omega}, dt) = \iint f(\underset{\sim}{\omega}, t)\, \underset{\sim}{\mu}_\lambda(d\underset{\sim}{\omega})\, dt.$$

Two homogeneous random measures λ and γ for which $\underset{\sim}{\mu}_\lambda = \underset{\sim}{\mu}_\gamma$ must be indistinguishable. Indeed, for $f(\underset{\sim}{\omega}, s) = 1_{(a,b]}(s)\, F \circ \underset{\sim}{\theta}_{-s}(\underset{\sim}{\omega})$, where $a < b \in \mathbb{R}$, $F \in \mathcal{F}^o(-\infty,\infty)$, (3.3) becomes

$$Q(F\, \lambda(a,b]) = \int f(\underset{\sim}{\omega}, s)\, \underset{\sim}{\mu}_\lambda(d\underset{\sim}{\omega})\, ds$$

$$= \int f(\underset{\sim}{\omega}, s)\, \underset{\sim}{\mu}_\gamma(d\underset{\sim}{\omega})\, ds = Q(F\, \gamma(a,b]).$$

The third measure ν_λ, corresponding to the Revuz-type measure on $\mathcal{E} \otimes \mathcal{E}$ considered by Sharpe in [17], was introduced in section 2. For $F \in b(\mathcal{E} \otimes \mathcal{E})^+$, ν_λ satisfies

(3.4) $$\int Q(d\underset{\sim}{\omega}) \int f(t, Z_{t-}(\underset{\sim}{\omega}), Z(\underset{\sim}{\omega}))\lambda(\underset{\sim}{\omega}, dt) = \iiint f(t,x,y)\, \nu_\lambda(dx,dy)\, dt,$$

when (A.3) and (H.2) hold, i.e., when the measure ν_λ defined by

(3.5) $$\iint f(x,y)\, \nu_\lambda(dx,dy) = Q \int_0^1 f(Z_{t-}, Z_t)\, \lambda(dt)$$

is σ-finite.

(3.6) PROPOSITION. *The measures* $\underset{\sim}{\mu}_\lambda$ *and* ν_λ *defined in* (3.2) *and* (3.5) *are* σ-*finite if* λ *is finite.*

PROOF. We prove the σ-finiteness of $\underset{\sim}{\mu}_\lambda$ and ν_λ at the same time by exhibiting a strictly positive random variable Y of the form $Y = F(Z_{0-}, Z_0) + 1_{\{Z_0 \notin E\}}$, where $F \in \mathcal{E} \otimes \mathcal{E}$ is strictly positive, for which $\underset{\sim}{\mu}_\lambda(Y) = \nu_\lambda(F) < \infty$. (Actually $F(x,y)$ will be a function of y alone.)

Fix $h > 0$ satisfying $\int h(x)\, \xi(dx) < \infty$, and consider first the case of λ "continuous" (not charging points). Define

$$(3.7) \qquad Y_c = \underset{\sim}{E}^{Z_0}[\int_0^\infty e^{-s}\, h(Z_s)\, e^{-\lambda(0,s]} ds]\, 1_{\{Z_0 \in E\}} + 1_{\{Z_0 \notin E\}}.$$

Then

$$\underset{\sim}{\mu}_\lambda(Y_c) = Q \int_0^1 \underset{\sim}{E}^{Z_t}\{\int_0^\infty e^{-s}\, h(Z_s)\, e^{-\lambda(0,s]} ds\}\, \lambda(dt)$$

$$= Q \int_0^1 e^t\, 1_{\{\underset{\sim}{\hat{\xi}} < t\}} \int_t^\infty e^{-u} h(Z_u)\, e^{-\lambda(t,u]} du\, \lambda(dt)$$

$$= \lim_n Q\{(\sum_{k=0}^{2^n} 1_{[a_{k-1,n}, a_{k,n})}(\underset{\sim}{\hat{\xi}})) \int_{k2^{-n}}^1 e^t\, e^{\lambda(k2^{-n},t]}$$

$$\int_t^\infty e^{-u} h(Z_u)\, e^{-\lambda(k2^{-n},u]} du\, \lambda(dt)$$

where $a_{k,n} = k2^{-n}$ for $k \geq 0$, $a_{-1,n} = -\infty$. After switching the order of integration in t and u (λ is Q-a.e. σ-finite) we obtain

$$\underset{\sim}{\mu}_\lambda(Y_c) \leq \lim_n Q\{\sum_{k=0}^{2^n} 1_{[a_{k-1,n}, a_{k,n})}(\hat{\xi})$$

$$\int_{k2^{-n}}^\infty e^{1-u}\, h(Z_u)\, e^{-\lambda(k2^{-n},u]}(e^{\lambda(k2^{-n}, u\wedge 1]} - 1) du\}$$

$$\leq Q\Big(e \int_0^\infty e^{-u}\, h(Z_u) du\Big) = e \int h(x)\, \xi(dx) < \infty.$$

The last equality follows from the way Q is defined.

Next suppose λ is "discrete" (purely atomic), with $\lambda\{t\}$ bounded, say by M. In this case define

$$m_s = \prod_{0<u\leq s} (M - \lambda\{u\})M^{-1}, \quad \text{and}$$

$$Y_{bd} = M^{-1} \underset{\sim}{E}^{Z_0}[\int_0^\infty e^{-s} h(Z_s)\, m_s\, ds]\, 1_{\{Z_0 \in E\}} + 1_{\{Z_0 \notin E\}}.$$

A calculation almost identical to the one above, using the identity

$$\sum_{0<t\leq u} \frac{\lambda\{t\}}{M\, m_t} = m_u^{-1} - 1$$

shows $\underset{\sim}{\mu}_\lambda(Y_{bd}) < \infty$.

If λ is discrete with $\lambda\{t\}$ finite, then for each $n \geq 1$

$$\lambda_n(s,t] = \sum_{s<u\leq t} \lambda\{u\}\, 1_{\{n-1 \leq \lambda\{u\} < n\}}$$

defines a homogeneous random measure having *bounded* atoms, and $\lambda = \sum_{n=1}^\infty \lambda_n$. The argument above proves associated measures $\underset{\sim}{\mu}_{\lambda_n}$ and ν_{λ_n} are σ-finite, and to conclude the same is true of $\underset{\sim}{\mu}_\lambda$ and ν_λ it suffices to prove that $\underset{\sim}{\mu}_{\lambda_n} \perp \underset{\sim}{\mu}_{\lambda_m}$ and $\nu_{\lambda_n} \perp \nu_{\lambda_m}$ if $m \neq n$. But this follows from the representations of section 1: for instance,

$$\begin{aligned} \underset{\sim}{\mu}_{\lambda_n}(A) &= Q \int_0^1 1_A \circ \underset{\sim}{\theta}_t\, \lambda_n(dt) \\ &= Q \int_0^1 1_{(A \cap \{f(Z_{0-}, Z_0) \in [n-1,n)\})} \circ \underset{\sim}{\theta}_t\, \lambda(dt) \end{aligned}$$

shows that the first pair of measures have disjoint support.

Now the proposition follows for general finite λ because, for Q-almost all $\underset{\sim}{\omega}$, λ has at most countably many finite atoms and therefore decomposes into the sum of a discrete measure λ_d and a non-atomic measure λ_c.

4. Application to Time Changing

In [12] it was shown that dual additive functional pairs embed in a single homogeneous random measure over Z. This correspondence between two additive functionals (one for each process) does not require *strong* duality hypotheses. The results of this section show that the pair of time changed processes arising from a pair of processes in *weak* duality and a pair of continuous additive functionals dual in the above sense will again be in weak duality. The duality measure is ν_λ, where λ is the measure over Z which embeds the additive functionals. This generalizes the known results for processes in strong duality.

Let λ be a continuous (diffuse), finite homogeneous random measure over Z, carried by $(\hat{\underset{\sim}{\zeta}},\underset{\sim}{\zeta})$ and define, for $p \in \mathbb{R}$,

$$(4.1) \qquad \underset{\sim}{\tau}_s^p = \begin{cases} \inf\{t > p\colon \lambda(p,t] > s\} & \text{if } s \geq 0, \\ \sup t < p\colon \lambda[t,p) \geq -s\} & \text{if } s < 0. \end{cases}$$

This is a right continuous strictly increasing process and for $s \geq 0$ $\underset{\sim}{\tau}_s^p$ is a stopping time, for $s < 0$, $\underset{\sim}{\tau}_s^p$ is a starting time. Note that

$$(4.2) \qquad \underset{\sim}{\tau}_s^{p+q} = \underset{\sim}{\tau}_s^p \circ \underset{\sim}{\theta}_q + q, \qquad Z(\underset{\sim}{\tau}_s^p) \circ \underset{\sim}{\theta}_u = Z(\underset{\sim}{\tau}_s^{p+u}).$$

When $p = 0$, write $\underset{\sim}{\tau}_s$ for $\underset{\sim}{\tau}_s^0$.

The main result of this section is the following.

(4.3) PROPOSITION. (i) $\underset{\sim}{\mu}_\lambda$ *is carried by* $\{Z_0 \in E\}$ *and* $\underset{\sim}{\tau}_0 = 0$ a.e. $\underset{\sim}{\mu}_\lambda$; (ii) *if* $\Gamma \subset \{\underset{\sim}{\omega}\colon Z(\underset{\sim}{\tau}_{-q}) \in E$ *and* $Z(\underset{\sim}{\tau}_0) \in E\}$, *then* $\underset{\sim}{\mu}_\lambda(\Gamma) = \underset{\sim}{\mu}_\lambda(\underset{\sim}{\theta}^{-1}_{\underset{\sim}{\tau}_q}(\Gamma))$.

PROOF. (i) This follows from the fact that λ is carried by

$(\hat{\zeta},\zeta) \cap \{u\colon \lambda(u,u+\varepsilon] > 0$ for all $\varepsilon > 0\}$.

(ii) First suppose Y is of the form

$$(4.4) \qquad Y(\omega) = \int_{-\infty}^{\infty} W(\omega,s)\, d\tau_s,$$

where W is a non-negative process and $d\tau_s$ is the Stieltjes measure generated by the increasing process $s \to \tau_s$. We will compute $\mu_\lambda(Y)$ using standard "time change" results in [2]. To facilitate this we introduce the (increasing) process

$$L_s = \begin{cases} \lambda(0,s] & \text{if } s \geq 0, \\ -\lambda[s,0) & \text{if } s < 0. \end{cases}$$

In terms of L, (4.4) becomes

$$(4.5) \qquad Y(\omega) = \int W(\omega,L_s(\omega))\, 1_{(-\infty,\infty)}(L_s(\omega))ds$$

$$= \int W(\omega,L_{-s}(\omega))\, 1_{(-\infty,\infty)}(L_{-s}(\omega))ds.$$

Using (4.5) and (3.3), and taking (i) into account,

$$(4.6) \quad \mu_\lambda(Y\, 1_{\{Z(\tau_{-q})\in E\}})$$

$$= Q\int W(\theta_s\omega,L_{-s}(\theta_s\omega))\, 1_{(-\infty,\infty)}(L_{-s}(\theta_s\omega))\, 1_{\{Z(\tau_{-q})\circ\theta_s\in E\}}\lambda(ds)$$

$$= Q\int W(\theta_s\omega,-L_s(\omega))\, 1_{\{Z(\tau_{-q})\circ\theta_s\in E\}}\, L(ds)$$

$$= Q\int W(\theta_{\tau_s(\omega)}(\omega),-s)\, 1_{\{Z(\tau_s)\in E,\ Z(\tau_{s-q})\in E\}}\, ds.$$

The last equality comes from a "time change"; the next-to-last follows from the identity $L_{-s}(\theta_s\omega) = -L_s(\omega)$.

For $\underset{\sim}{\omega} \in \{\hat{\underset{\sim}{\zeta}} < \underset{\sim}{\tau}_q < \underset{\sim}{\zeta}\}$ we have

$$\underset{\sim}{\tau}_{s+q}(\underset{\sim}{\omega}) = \underset{\sim}{\tau}_q(\underset{\sim}{\omega}) + \underset{\sim}{\tau}_s \circ \underset{\sim}{\theta}_{\underset{\sim}{\tau}_q}(\underset{\sim}{\omega}),$$

and

$$\begin{aligned} Y \circ \underset{\sim}{\theta}_{\underset{\sim}{\tau}_q}(\underset{\sim}{\omega}) &= [\int W_s \, d_{\underset{\sim}{\tau}_s}] \circ \underset{\sim}{\theta}_q(\underset{\sim}{\omega}) \\ &= \int W(\underset{\sim}{\theta}_{\underset{\sim}{\tau}_q}(\underset{\sim}{\omega})\underset{\sim}{\omega}, s - q) d\underset{\sim}{\tau}_s \\ &= \int W(\underset{\sim}{\theta}_{\underset{\sim}{\tau}_q}(\underset{\sim}{\omega})\underset{\sim}{\omega}, L_s(\underset{\sim}{\omega}) - q) \, 1_{(-\infty,\infty)}(L_s(\underset{\sim}{\omega}))ds \\ &= \int W(\underset{\sim}{\theta}_{\underset{\sim}{\tau}_q}(\underset{\sim}{\omega})\underset{\sim}{\omega}, L_{-s}(\underset{\sim}{\omega}) - q) 1_{(-\infty,\infty)}(L_{-s}(\underset{\sim}{\omega}))ds. \end{aligned}$$

The same steps which led to (3.6) above yield

$$\begin{aligned} \underset{\sim}{\mu}_\lambda(Y \circ \underset{\sim}{\theta}_{\underset{\sim}{\tau}_q} 1_{\{Z(\underset{\sim}{\tau}_q) \in E\}}) &= Q \int W(\underset{\sim}{\theta}_{\underset{\sim}{\tau}_q} \circ \underset{\sim}{\theta}_s \underset{\sim}{\omega}, L_{-s} \circ \underset{\sim}{\theta}_s \underset{\sim}{\omega} - q) 1_{\{Z(\underset{\sim}{\tau}_q) \circ \underset{\sim}{\theta}_s \in E\}} \lambda(ds) \\ &= Q \int W(\underset{\sim}{\theta}_{\underset{\sim}{\tau}_{q+s}} \underset{\sim}{\omega}, -s - q) \, 1_{\{Z(\underset{\sim}{\tau}_{q+s}(\underset{\sim}{\omega})) \in E\}} \, 1_{\{Z(\underset{\sim}{\tau}_s(\underset{\sim}{\omega})) \in E\}} \, ds \\ &= Q \int W(\underset{\sim}{\theta}_{\underset{\sim}{\tau}_t} \underset{\sim}{\omega}, -t) \, 1_{\{Z(\underset{\sim}{\tau}_t) \in E\}} \, 1_{\{Z(\underset{\sim}{\tau}_{t-q}) \in E\}} \, ds \\ &= \underset{\sim}{\mu}_\lambda(Y \, 1_{\{Z(\underset{\sim}{\tau}_{-q}) \in E\}}). \end{aligned}$$

To verify (ii) we chose the process W in (4.4) so that $Y = 1_\Gamma$. Since by assumption if $\underset{\sim}{\omega} \in \Gamma$ we must have $0 < |\underset{\sim}{\tau}_0(\underset{\sim}{\omega}) - \underset{\sim}{\tau}_{-q}(\underset{\sim}{\omega})| < \infty$, the process

$$\begin{aligned} W(s) &= |\underset{\sim}{\tau}_0 - \underset{\sim}{\tau}_{-q}|^{-1} \cdot 1_\Gamma \quad \text{for } s \text{ between } 0 \text{ and } -q \\ &= 0 \quad \text{elsewhere} \end{aligned}$$

satisfies this condition. With this choice

$$\underset{\sim}{\mu}_\lambda(\Gamma) = \underset{\sim}{\mu}_\lambda(Y\, 1_{\{Z(\underset{\sim}{\tau}_{-q}) \in E\}})$$

$$= \underset{\sim}{\mu}_\lambda(Y \cdot 1_{\{Z(\underset{\sim}{\tau}_0) \in E\}} \circ \underset{\sim}{\theta}_{\underset{\sim}{\tau}_q})$$

$$= \underset{\sim}{\mu}_\lambda(\underset{\sim}{\theta}^{-1}_{\underset{\sim}{\tau}_q} \Gamma).$$

(4.7) COROLLARY. *Suppose* λ *corresponds to the dual pair* $A, \hat{A}$ *of finite continuous additive functionals of* $X, \hat{X}$. *Let* τ_s, $\hat{\tau}_s$ *be the corresponding inverses. Then for* $A, B, C \in \mathcal{E}$ *and* $s, t > 0$, *the following are equal:*

$$\int_B P^x(X_{\tau_t} \in C)\, \hat{P}^x(\hat{X}_{\hat{\tau}_s} \in A)\, \nu_\lambda(dx)$$

$$\int_A P^x(X_{\tau_s} \in B,\ X_{\tau_{t+s}} \in C)\, \nu_\lambda(dx)$$

$$\int_C \hat{P}^x(\hat{X}_{\hat{\tau}_t} \in B,\ \hat{X}_{\hat{\tau}_{t+s}} \in A)\, \nu_\lambda(dx).$$

PROOF. Define $\Gamma = \{\underset{\sim}{\omega}\colon Z(\underset{\sim}{\tau}_{-s}) \in A,\ Z(\underset{\sim}{\tau}_0) \in B,\ Z(\underset{\sim}{\tau}_t) \in C\}$. By (4.3)(ii) $\underset{\sim}{\mu}_\lambda(\Gamma) = \underset{\sim}{\mu}_\lambda(\underset{\sim}{\theta}^{-1}_{\underset{\sim}{\tau}_q} \Gamma)$ for $-t \le q \le s$. Computing $\underset{\sim}{\mu}_\lambda(\Gamma)$ and the right-hand side for $q = -t$ and $q = s$, we obtain three expressions whose equality will imply the result:

$$\underset{\sim}{\mu}_\lambda(\Gamma) = Q \int_0^1 1_A(Z(\underset{\sim}{\tau}_{-s}) \circ \underset{\sim}{\theta}_u)\, 1_B(Z_u)\, 1_C(Z(\underset{\sim}{\tau}_t) \circ \underset{\sim}{\theta}_u)\, \lambda(du)$$

$$= Q \int_0^1 \hat{\underset{\sim}{P}}^{Z_{u-}}(Z(\underset{\sim}{\tau}_{-s}) \in A)\, \underset{\sim}{P}^{Z_u}(Z(\underset{\sim}{\tau}_t) \in C)\, 1_B(Z_u)\, \lambda(du)$$

$$= \int_B \hat{P}^x(Z(\underset{\sim}{\tau}_{-s}) \in A)\, P^x(Z(\underset{\sim}{\tau}_t) \in C)\, \nu_\lambda(dx).$$

To get the second equality we used the fact that λ is both optional and cooptional.

Similarly,

$$\underset{\sim}{\mu}_\lambda(\theta^{-1}_{\underset{\sim}{\tau}_s}\Gamma) = \int_A \underset{\sim}{P}^x(Z(\underset{\sim}{\tau}_s) \in B,\ Z(\underset{\sim}{\tau}_{s+t}) \in C)\ \nu_\lambda(dx)$$

$$\underset{\sim}{\mu}_\lambda(\theta^{-1}_{\underset{\sim}{\tau}_{-t}}\Gamma) = \int_C \hat{\underset{\sim}{P}}^x(Z(\underset{\sim}{\tau}_{-t}) \in B,\ Z(\underset{\sim}{\tau}_{-t-s}) \in A)\ \nu_\lambda(dx).$$

Now, $\underset{\sim}{P}^x(Z(\underset{\sim}{\tau}_s) \in B,\ Z(\underset{\sim}{\tau}_{t+s}) \in C) = P^x(X_{\tau_s} \in B,\ X_{\tau_{t+s}} \in C)$ is a straightforward consequence of the relationships between Z and X, λ and A, $\underset{\sim}{P}^x$ and P^x (see [12] and [13]). The expressions involving $\hat{\underset{\sim}{P}}^x$ transform into versions involving the *left* limits of the process $t \to \hat{X}_{\hat{\tau}_t}$. This process has at most countably many discontinuities, and ν_λ is excessive relative to this process (see [14], III.3). Using these two facts, a simple argument shows that the process must be $\hat{P}^{\nu_\lambda}$ - a.s. continuous at any fixed time, so that we may drop the left limits in the transformed expressions. Thus the announced equalities hold.

References

1. C. DELLACHERIE. Ensembles aléatoires I, II. *Séminaire de Probabilités III (Univ. Strasbourg)*, pp. 97-136. Lecture Notes in Math. *88*, Springer-Verlag, Berlin, 1967.

2. C. DELLACHERIE. *Capacités et Processus Stochastiques.* Springer-Verlag, New York, 1972.

3. E.B. DYNKIN. Additive functionals of Markov processes and stochastic systems. *Ann. Inst. Fourier (Grenoble) 25* (1975), 177-200.

4. E.B. DYNKIN. Additive functionals of several time-reversible Markov processes. *J. Functional Analysis 42* (1981), 64-101.

5. E.B. DYNKIN. Green's and Dirichlet spaces associated with fine Markov processes. (Preprint, 1982)

6. E.B. DYNKIN. Markov systems and their additive functionals. *Ann. Probab. 5* (1977), 653-677.

7. M. FUKUSHIMA. *Dirichlet Forms and Markov Processes.* North-Holland, New York/Kodansha, Tokyo, 1980.

8. D. GEMAN and J. HOROWITZ. Remarks on Palm measures. *Ann. Inst. Henri Poincaré, Sec. B. IX* (1973), 215-232.

9. R.K. GETOOR. Duality of Lévy systems. *Z. Wahrscheinlichkeitstheorie verw. Gebiete 19* (1971), 257-270.

10. P.A. MEYER. Ensembles aléatoires Markoviens homogènes I, *Séminaire de Probabilités III (Univ. Strasbourg)*, pp. 176-190. Lecture Notes in Math. *381*, Springer-Verlag, Berlin, 1974.

11. J.B. MITRO. Balayage and exit systems for dual Markov processes. (Preprint, 1982)

12. J.B. MITRO. Dual Markov functionals: applications of a useful auxiliary process. *Z. Wahrscheinlichkeitstheorie verw. Gebiete 48* (1979), 97-114.

13. J.B. MITRO. Dual Markov processes: construction of a useful auxiliary process. *Z. Wahrscheinlichkeitstheorie verw. Gebiete 47* (1979), 139-156.

14. D. REVUZ. Measures associées aux fonctionelles additives de Markov I. *Trans. Amer. Math. Soc. 148* (1970), 501-531.

15. D. REVUZ. Measures associées aux fonctionelles additives de Markov II. *Z. Wahrscheinlichkeitstheorie verw. Gebiete 16* (1970), 336-344.

16. J. de SAM LAZARO and P.A. MEYER. Hélices croissantes et mesures de Palm. *Séminaire de Probabilités IX (Univ. Strasbourg)*, pp. 38-51. Lecture notes in Math. *465*, Springer-Verlag, Berlin, 1975.

17. M.J. SHARPE. Discontinuous additive functionals of dual processes. *Z. Wahrscheinlichkeitstheorie verw. Gebiete 21* (1975), 81-95.

B.W. ATKINSON
Department of Mathematics
University of Southern California
Los Angeles, CA 90089
Current Address:
Department of Mathematics
University of Florida
Gainesville, FL 32611

J.B. MITRO
Department of
Mathematical Sciences
University of Cincinnati
Cincinnati, Ohio 45221

Seminar on Stochastic Processes, 1982
Birkhäuser, Boston, 1983

OCCUPATION TIMES OF d-DIMENSIONAL SEMIMARTINGALES

by

R. F. BASS

1.0 Introduction

If X_t is a stochastic process, f a nonnegative function, define $A_t(f) = \int_0^t f(X_s)ds$. When X_t is a Markov process, $A_t(f)$ is an example of what are called additive functionals. Over the past 20 years, an extensive theory has been developed about additive functionals of Markov processes. One of the major results is that, given a measure μ, under quite general conditions on X and μ, one can construct an additive functional $A_t(\mu)$ as the limit of additive functionals $A_t(f_n)$ for some sequence of functions f_n. When μ is point mass at a point x, $A_t(\mu)$ is called the local time at x. More generally, when μ is concentrated on a set C, $A_t(\mu)$ is called (one of) the local times at the set C. Local times have turned out to be among the most useful examples of additive functionals.

Suppose now that X_t is a d-dimensional semimartingale: an $\mathbb{R}^d$-valued process, each of whose coordinates is a semimartingale. Let us call processes $A_t(f)$ defined above, as well as uniform limits of such processes, occupation times. (The reason for the change of name is that there will, in general, be processes that satisfy the abstract definition of additive functional but are not occupation times.)

Our goals in this paper are

(1.1) to construct an occupation time $A_t(\mu)$ corresponding to a measure μ;

(1.2) to give conditions under which $A_t(\mu_a)$ is jointly continuous in a and t;

(1.3) to give integral representations of $A_t(\mu)$ and $A_t(f)$ in terms of $A_t(\lambda(s,v))$, where $\lambda(s,v)$ is $d-1$ dimensional Lebesgue measure on the hyperplane $\{y: y*v = s\}$, v on the unit ball, $y*v$ the inner product of y and v.

Of these, (1.3) is the generalization to the d-dimensional semimartingale case of the results obtained in [1] for Brownian motion.

The precise results are given in section 2 as theorems T1, T2, T3 and T4. Let us briefly describe the hypotheses needed on X_t and μ (given more precisely in section 2 as A2 and M1). The process X_t must satisfy $dX_t = \sigma_t(X)dW_t + \eta_t\,dt$. Here η_t need only be bounded and predictable; $\sigma_t(X)$ is bounded, strictly elliptic, continuous in t, and for each t, $\sigma_t(X)$ is a functional on the path $\{X_s, s \le t\}$. Note that $\sigma_t(X)$ is a functional of the entire past of X, not just the current value X_t as is the case when X is Markov. As a functional, we would like σ_t to be twice Fréchet differentiable; in fact, slightly less will do.

The conditions we put on μ are that $\mu(\mathbb{R}^d)$ be finite (this can be weakened: see 6.1) and that $\mu(B(r,x)) \le cr^{d-2+\varepsilon}$, where $B(r,x)$ is the ball of radius r centered at x and c,ε are constants > 0 independent of r and x. If μ has a bounded density, we could let $\varepsilon = 2$; if μ is concentrated on a surface in a nice way, we could let $\varepsilon = 1$. Of course, μ could be much wilder than either of these (cf. 8.3).

Looking at the particular case when X_t is Markov, (1.1) is, as we stated above, known; (1.3) is new, while (1.2), at least in the generality in which it is stated, may perhaps be new. At the heart of our method are estimates on the size and smoothness of $\rho(z)$, the density of $\lambda(C) = E\int_0^t 1_C(X_s)ds$ with respect to Lebesgue measure. Bounds on $\rho(z)$ are well-known when σ_t is only strictly elliptic and Hölder continuous. However, to get smoothness of $\rho(z)$, one needs that the coefficients of the adjoint operator for X be Hölder continuous, which translates to requiring that σ be in $C^{2+\alpha}$ for some α (see [3, Chapter 6]). Thus, even in the Markov case, the classical results on parabolic partial differential equations do not give an improvement on our results for (1.2) and (1.3).

Returning to the general case, Maisonneuve [7] has a way of constructing an increasing process which has support $\{(t,\omega): X_t(\omega) \in C\}$. Such a process need not be an occupation time, however. Krylov [6] has bounds on the size of the density $\rho(z)$, but his bounds are not good enough for our purposes and his methods give no information about the smoothness of ρ. If one did not want (1.1), (1.2), or (1.3), but only wanted to construct some occupation time on a very smooth surface C, there is an easy way to proceed. One maps $\mathbb{R}^d$ into $\mathbb{R}^d$ smoothly so that C gets mapped into a subset of $\mathbb{R}^{d-1}\times\{0\}$, and one now has what is essentially a one-dimensional problem. This procedure is carried out in section 8, but it works only when C is quite smooth.

Since our argument is rather long, we will give a brief outline. In section 2 we introduce some notation and give statements of the results. First proving our results under the restrictions that σ_t is close to the identity matrix I, that $\eta_t = 0$, and that the dimension $d \geq 3$ (see R1,2,3 of section 2), we begin in section 3 with an integration by parts formula. We take a formula of Haussmann [4], and following the approach used by Bismut [2] to prove Malliavin's results, we apply it to the functional $L(X) = \int_0^\tau g(X_s)h_s ds$, where τ is a fixed time, g a differentiable function, and h_s functionals to be specified later. After some calculations, we get that

$$E \int_0^\tau Dg(X_s)G_s h_s ds$$

is equal to an expression involving $g(X_s)$, h_s, and some other terms (see 3.4), where G_s is a certain auxiliary process and Dg is the gradient of g. We would like to let $h_s = G_s^{-1}$ and then use Hölder's inequality. But, easy examples show that G_s^{-1} need not be integrable except in the Markov case. This is not just a technical point: it turns out that G_s is intimately related to the Fréchet derivative of the mapping $W_t \to X_t$. The singularity of G has implications to the existence of densities for X_t. The device we use to get around the singularity of G is to stop at a randomized stopping time; the randomization is necessary to preserve Fréchet differentiability at appropriate places.

After using Hölder's inequality, we are faced with bounding expressions of the form $E\int_0^\tau |g|^p(X_s)s^{-\beta}ds$, $0 < \beta < \frac{1}{2}$. We do this in section 4 by comparing the radial process $|X_s - w|$ to a Bessel process of appropriate index ν. (This is why we need $\sigma_t \approx I$.) Letting g equal successively the first order partial derivatives of the Newtonian potential of the indicator of $B(r,w)$, we achieve the necessary estimates for the density of the potential of the stopped process.

In section 5 we use these estimates to get estimates of $|E \int_0^{S\wedge\tau}(\mu * \phi_b - \mu * \phi_a)(X_s)ds|$, S our stopping time, ϕ_b an approximation to the identity. We do an iteration scheme to suppress S from this expression, and then, following roughly along the lines of the corresponding proof for Markov additive functionals, we estimate $E|\int_0^\tau(\mu * \phi_b - \mu * \phi_a)(X_s)ds|^p$.

Using Kolmogorov's theorem and techniques from [1], we prove our results in section 6, subject to the restrictions R1,2,3. Finally, in section 7, we eliminate these restrictions: $\sigma_t \approx I$ by an iteration procedure, no drift by a Girsanov transformation, $d \geq 3$ by a projection argument.

2.0 Notation and statement of results

If $x \in \mathbb{R}^d$, $|x|$ is the Euclidean norm of x. If K is a $d \times d$ matrix, $|K| = \sup\{|Kx|: |x| \leq 1\}$ and K^* is the transpose of K. e_j will denote the jth standard basis vector for $\mathbb{R}^d$. Let $B(r,x) = \{y \in \mathbb{R}^d: |y - x| < r\}$.

If f is an $\mathbb{R}$ or $\mathbb{R}^d$-valued function, $\|f\| = \sup\{|f(x)|\}$; if the domain of f is $[0,\infty)$, $\|f\|_\tau = \sup\{|f(t)|: t \leq \tau\}$; let

$$\|f\|_{C^\alpha} = \|f\| + \sup_{y \neq x}|f(y) - f(x)|/|y - x|^\alpha.$$

If μ is a measure, $\|\mu\|_V$ is the total variation of μ and $\|\mu\|_{W-\alpha} = \sup\{|\int f d\mu| : \|f\|_{C^\alpha} \leq 1\}$. If μ is a measure on $[0,\infty)$, we will let $[\mu,f]$ denote $\int\mu(ds)f(s)$; we will frequently write integrals in this order since we identify elements of $\mathbb{R}^d$ with $d \times 1$ matrices, with the exception that we will consider $\mathbb{R}^d$-valued measures as $1 \times d$ matrices.

If L is a real-valued functional on $C[0,\infty)$, the continuous $\mathbb{R}^d$-valued functions, the Fréchet derivative of L at f is an $\mathbb{R}^d$-valued measure $L'(f)$ such that

$$|L(f + g) - L(f) - [L'(f),g]| = o(\|g\|).$$

The letter c with a subscript will denote constants; $c_{2.1}(\alpha,n)$, for example, is a constant whose value depends on α and n.

Let (Ω,F,P) be a probability space. When necessary, we will assume $\Omega = C[0,\infty)$; let F_t be the right continuous, completed filtration generated by W_t, a d-dimensional Brownian motion on Ω. Let W_t^j be the jth coordinate of W_t. If M is a martingale, let $\langle M,M\rangle$ be the quadratic variation of M (see [8] for details about stochastic integration). Let $\phi_b(y) = (2\pi b)^{-d/2} e^{-|y|^2/2b}$.

Suppose X_t satisfies

$$X_0 = x_0, \quad dX_t = \sigma_t dW_t + \eta_t dt, \tag{2.1}$$

where $\sigma_t \in \mathbb{R}^{d\times d}$, $\eta_t \in \mathbb{R}^d$, respectively, and are both bounded and predictable. We will refer to σ_t and η_t as the diffusion and drift coefficients of X_t respectively. We will suppose $\sigma_t(\omega) = \hat{\sigma}_t(X(\omega))$, a.s., where $\hat{\sigma}_t$ is a $d\times d$ matrix valued functional on $C[0,t]$. Since no confusion should result, we will drop the ^'s on the $\hat{\sigma}$ and write σ_t, $\sigma_t(X)$, $\sigma(t)$, or $\sigma(t,X)$ as needed.

We will make the following assumptions on σ and η.

A1 (i) $|\eta_t|$ *is bounded by* $c_{2.1}$, *independent of* t; η_t *is predictable;*

(ii) $|\sigma_t|$ *is bounded by* $c_{2.2}$, *independent of* t; σ_t *is predictable;*

(iii) σ_t *is uniformly strictly elliptic: there exists* $c_{2.3}$ *independent of* t *such that* $x^*\sigma_t x \ge c_{2.3}|x|^2$ *for all* x;

(iv) *for each* i,j, $\sigma_{ij}(t)$ *is a functional on* $C[0,t]$ *which is Fréchet differentiable;* $\sup_{t\le\tau}|\,\|\sigma'_{ij}(t,f)\|_V| \le c_{2.4}(\tau)$, $c_{2.4}$ *independent of* f, i, *and* j;

(v) $\sigma'_{ij}(t,f)$ *is itself Fréchet differentiable: there exists*

$\sigma''_{ij}(t)$, *a measure-valued linear functional on* $C[0,t]$ *such that* $|\|\sigma'_{ij}(t,f+g) - \sigma'_{ij}(t,f) - \sigma''_{ij}(t,f)(g)\|_V| = o(\|g\|_t)$; *moreover,* $\sup_{t\le\tau}|\|\sigma''_{ij}(t,f)(g)\|_V| \le c_{2.5}(\tau)\|g\|_\tau$, $c_{2.5}$ *independent of* f, i, *and* j.

In section 5, we will weaken A1 to:

A2 (i) A1(i) *still holds;*

(ii) *there exists a sequence of functionals* σ^n *satisfying* A1(ii)-(v), *the constants* $c_{2.2}, c_{2.3}, c_{2.4}, c_{2.5}$ *independent of* n, *such that for each* $f \in C[0,\tau]$, $\sup_{t\le\tau}|\sigma^n_t(f) - \sigma_t(f)| \to 0$.

We impose the following condition on measures μ (and μ_a and λ, where appropriate)

M1 (i) $\mu(\mathbb{R}^d) \le c_{2.6} < \infty$;

(ii) *for all* x *and* r, $\mu(B(r,x)) \le c_{2.7}r^{d-2+c_{2.8}}$, *where* $c_{2.7}, c_{2.8}$ *are independent of* x,r *and are* > 0.

We can now state our main theorems: T1,2,3 and 4. In each, we suppose X is given by 2.1 and we suppose A2 holds. Let $A_t(f) = \int_0^t f(X_s)ds$. Let $\mu * \phi_b(x)$ be the density of the measure $\mu * \phi_b(dx)$.

T1 THEOREM. *Suppose* μ *satisfies* M1. *Then there exists an increasing, continuous process* $A_t(\mu)$ *such that for all* τ,

$$\sup_{t\le\tau}|A_t(\mu * \phi_b) - A_t(\mu)| \to 0, \text{ a.s. } \textit{as } b \to 0.$$

T2 THEOREM. *Suppose* $\{\mu_a, 0 \le a \le 1\}$ *satisfies* M1 *for each* a, $c_{2.6}$, $c_{2.7}$, $c_{2.8}$ *independent of* a. *Suppose for some* α,

$$\|\mu_a - \mu_b\|_{W-\alpha} \le c_{2.9}(\alpha)|b - a|^{c_{2.10}(\alpha)}.$$

Then there exist versions of $A_t(\mu_a)$ *such that*

$$\sup_{t\le\tau}\ \sup_{|a-b|<\delta} |A_t(\mu_a) - A_t(\mu_b)| \to 0, \quad \text{a.s., } as\ \delta \to 0.$$

Let

$$I_b(y) = (2\pi)^{-d}\int_0^\infty \cos(qy)q^{d-1}e^{-bq^2/2}\,dq, \tag{2.2}$$

and

$$\hat{B}_t(\mu,b) = \int \int_{(|v|=1)} \int_{-\infty}^{\infty} I_b(s - y^*v)A_t(\lambda(s,v))ds\,dv\,\mu(dy). \tag{2.3}$$

Here the integral with respect to dv is with respect to surface measure on the unit ball, and $\lambda(s,v)$ is $d-1$ dimensional Lebesgue measure on the hyperplane $\{z\colon z^*v = s\}$.

T3 THEOREM. *Suppose* μ *satisfies* M1. *Then*

$$\sup_{t\le\tau}|A_t(\mu) - \hat{B}_t(\mu,b)| \to 0, \quad \text{a.s. } as\ b \to 0.$$

If f is a continuous function on $\mathbb{R}^d$, let

$$g_v(s) = \int_{(y^*v=s)} f(y)d\lambda(s,v).$$

T4 THEOREM. *Suppose* μ *satisfies* M1. *Suppose* d *is odd and* $D_{d-1}g_v(s) = \partial^{d-1}g_v(s)/\partial s^{d-1}$ *is in* $L_1(-\infty,\infty)$. *Then*

$$A_t(f) = \tfrac{1}{2}(-1)^{(d-1)/2}\int_{(|v|=1)}\int_{-\infty}^{\infty} D_{d-1}g_v(s)A_t(\lambda(s,v))ds\,dv, \text{ a.s.}$$

T1,2,3, and 4 will be first proved under some restrictions; the restrictions will be removed in section 7.

R1 *There exists a constant* $c_{2.11} < (d+1)^{-1}(c_{2.8} \wedge 1)/16$ *such that* $|\sigma_t - I| \le c_{2.11}$ *for all* t.

R2 $\eta_t \equiv 0$.

R3 $d \ge 3$.

Finally, we need to state Haussmann's formula. Since we will be considering functionals that will be defined in terms of stochastic integrals, we need to modify slightly what we mean by the Fréchet derivative of such a functional.

Let S be the set of $C[0,\infty)$-valued processes adapted to F_t, identifying two processes that are equal, P-a.s. If L is a functional defined on S, mapping Z to an r.v. L(Z), we will say L is P-Fréchet differentiable if L satisfies

(2.4) (i) $E|L(Z)|^q \le c_{2.12}(q,Z) < \infty$ for all q;

(ii) $L(Z) \in F_\tau$ for some τ independent of Z;

(iii) if Q is any probability equivalent to P, if $Z, Y \in S$, and if the P-distribution of Z is equal to the Q-distribution of Y, then the P-distribution of L(Z) is equal to the Q-distribution of L(Y);

(iv) there exists a functional L' mapping $Z \in S$ to L'(Z), a random measure (i.e., for each ω, L'(Z) is a measure on $C[0,\tau]$) such that $E|L(Z+Y) - L(Z) - [L'(Z),Y]|^q = o(E\|Y\|^q)$ for all q;

(v) $E|\|L'(Z)\|_V|^q \le c_{2.13}(q) < \infty$.

We define the auxiliary $\mathbb{R}^{d\times d}$-valued process ψ_t^s by

(2.5) $\psi_t^s = 0$ if $t < s$; $\psi_t^s e_j = Ie_j + \int_s^t a(s,r)dW_r$ if $t \geq s$, $1 \leq j \leq d$,

where $a(s,r)$ is $\mathbb{R}^{d\times d}$-valued and $a(s,r)_{ik} = [\sigma'_{ik}(r), \psi_\cdot^s e_j]$.

We then have, upon making trivial modifications to his proof, Haussmann's formula

(2.6) THEOREM (Haussmann [4]). *If* L *satisfies* (2.4), σ *satisfies* A1, X *is given by* (2.1), *and* R2 *holds, then*

$$L(X) = EL(X) + \int_0^\tau E([L'(X),\psi_\cdot^t]|F_t)\sigma_t dW_t.$$

3.0 Integration by parts

Throughout this section we assume that σ satisfies A1 and that R1,2,3 hold. We need to introduce some auxiliary processes.

Let G_t, J_t be $\mathbb{R}^{d\times d}$-valued processes given by

$$G_t = \int_0^t \psi_t^s ds, \quad J_t = t^{-1}G_t,$$

where ψ_t^s is given by (2.5).

We need the following facts about ψ, G, and J.

(3.1) LEMMA. (i) $E \sup_{s\leq r\leq t} |\psi_r^s|^q \leq c_{3.1}(q,t)$;

(ii) $E \sup_{s\leq t} |G_s - sI|^q \leq c_{3.2}(q,t)$;

(iii) $J_t \to I$, a.s., *as* $t \to 0$; $E \sup_{s\leq t} |J_s - I|^q \leq c_{3.3}(q,t)$;

(iv) *If* $\theta > 1/8$ *and* $S_\theta = \inf\{t: |J_t - I| > \theta\}$, *then*
$E\, e^{-S_\theta} < c_{3.4}(\theta) < 1$.

PROOF. We will omit the proof (i) since it is similar to (ii), but easier.

(ii) Let $b_{ik}(r) = \int_0^r a_{ik}(s,r)ds = \int_0^r \int_0^r \sigma'_{ik}(r)(du)\psi_u^s e_j ds$

$$= [\sigma'_{ik}(r), G e_j],$$

using Fubini and the fact that $\psi_u^s = 0$ if $u < s$. It follows then that

$$|b_{ik}(r)| \le \|\sigma'_{ik}(r)\|_V \sup_{u\le r}|G_u e_j| \le c_{3.5} \sup_{u\le r}|G_u e_j|, \quad c_{3.5} = c_{2.4}(r).$$

Suppose $H(s,r) = 1_{[s_1,s_2]}(s)1_F(\omega)1_{[r_1,r_2]}(r)$, where $s_1 \le s_2 \le r_1 \le r_2$, $F \in \mathcal{F}_{r_1}$. Direct calculation shows that

$$\int_0^t \int_0^t H(s,r)dW_r ds = \int_0^t \int_0^t H(s,r)ds dW_r. \tag{3.2}$$

By linearity and taking limits, (3.2) holds for $H(s,r) = a_{ik}(s,r)1_{(s\le r)}$, and so we get

$$G_t e_j = \int_0^t \psi_t^s e_j ds = te_j + \int_0^t \int_s^t a(s,r)dW_r ds = te_j + \int_0^t b_r dW_r.$$

Each component of $G_t e_j - te_j$ is a martingale, and so, using Burkholder's inequality and taking $q \ge 2$,

$$E \sup_{u\le t}|G_u e_j - ue_j|^q \le c_{3.6}(q)E\Big(\int_0^t \operatorname{trace}(b_r^* b_r)dr\Big)^{q/2}$$

$$\le c_{3.7}(q)E\Big(\int_0^t \sup_{u\le r}|G_u e_j|^2 dr\Big)^{q/2}$$

$$\le c_{3.8}(q)t^{q/2-1}\, E\int_0^t \Big(\sup_{u\le r}|G_u e_j - ue_j|^q + r^q\Big)dr \quad \text{(Hölder).}$$

By Gronwall's inequality, (ii) follows with $c_{3.2}(q,t) = c_{3.9}(q)t^{3q/2}$ for $t \le 1$.

$$\text{(iii)}\quad E \sup_{u \le 2^{-p}} |J_u - I|^q \le E \sum_{i=p}^{\infty} \sup_{2^{-(i+1)} \le u \le 2^{-i}} |J_u - I|^q$$

$$\le E \sum_{i=p}^{\infty} \sup_{2^{-(i+1)} \le u \le 2^{-i}} |G_u - uI|^q \, 2^{(i+1)q}$$

$$\le c_{3.9}(q) \sum_{i=p}^{\infty} 2^{(i+1)q}(2^{-i})^{3q/2} \le c_{3.10}(q)2^{-p},$$

if $q \ge 2$. This proves that $J_t \to I$ a.s. as $t \to 0$, and the fact that $|J_u - I| \le 2|G_u - uI|$ if $u \ge \frac{1}{2}$ gives the remainder of (iii).

$$\text{(iv)}\quad P(S_\theta \le t) = P(\sup_{u \le t}|J_u - I| \ge \theta) \le \theta^{-2}E(\sup_{u \le t}|J_u - I|^2)$$

$$\le c_{3.10}(2)\theta^{-2}2^{-p}$$

if $t = 2^{-p}$ for some p; here $c_{3.10}$ depends only on $c_{2.4}(1)$. Since $\theta > 1/8$, take p sufficiently large so that $P(S_\theta \le 2^{-p}) \le \frac{1}{2}$. Then $Ee^{-S_\theta} \le \frac{1}{2} + \frac{1}{2}e^{-2^{-p}} = c_{3.11} < 1$. □

(3.3) LEMMA. ψ_t^s, G_t, J_t, $t \le \tau$ *each satisfy* (2.4 i - v); *in fact, the constant* $c_{2.13}$ *in* (2.4v) *may be chosen so that*

$$E \sup_{t \le \tau}|\|\psi_t^{s\prime}\|_V|^q,\ E \sup_{t \le \tau}|\|G_t'\|_V|^q,\ \textit{and}\ E \sup_{t \le \tau}|\|J_t'\|_V|^q\ \textit{are all} \le c_{2.13}.$$

PROOF. The proof for G_t is similar to that for ψ_t^s, while the results for J_t follow from those for G_t as in (3.1). Thus we will prove only the results for ψ_t^s. (i) and (ii) follow from (3.1) and the definitions.

(iii) $$d\psi_t^s e_j = a(s,t)dW_t = a(s,t)\sigma_t^{-1}dX_t,$$

where $a(s,t)$ depends on ψ_t^s and X. This stochastic differential equation may be solved by Picard iteration, from which (iii) follows.

(iv) Let $\psi_t^s(X)' = 0$ if $t < s$, and solve, for each j and each continuous process Y,

$$[\psi_t^s(X)'e_j,Y] = \int_s^t a'(s,r,X,Y)dW_r \quad \text{if } t \geq s,$$

where

$$a_{ik}'(s,r,X,Y) = [\sigma_{ik}'(r,X),[\psi_\cdot^s(X)'e_j,Y]] + [\sigma_{ik}''(r,X)(Y),\psi_\cdot^s e_j].$$

We claim that $\psi_t^s(X)'$ is the required random measure of (2.4iv). The proof is so similar to steps in the proof of Haussmann's formula (cf. [4, p.23] and also the proof (3.1ii) above, essentially applications of Burkholder's and Gronwall's inequalities, that we omit the details.

(v) $$|a_{ik}'(s,r,X,Y)| \leq \|\sigma_{ik}'(r,X)\|_V \sup_{u\leq r}|[\psi_u^s(X)'e_j,Y]| + \|\sigma_{ik}''(r,X)(Y)\|_V \sup_{u\leq r}|\psi_u^s e_j|.$$

By Burkholder's inequality, if $\|Y\|_\tau \leq 1$, a.s.,

$$E\sup_{u\leq t}[\psi_u^s(X)'e_j,Y]|^q \leq c_{3.12}(q,\tau)E\Big(\int_0^t \mathrm{trace}(a'(s,r,X,Y)^*a'(s,r,X,Y))dr\Big)^{\frac{q}{2}}$$

$$\leq c_{3.13}(q,\tau)E\int_0^t \sup_{u\leq r}|[\psi_u^s(X)'e_j,Y]|^q dr + c_{3.14}(q,\tau)E\int_0^t \sup_{u\leq r}|\psi_u^s|^q dr\,\|Y\|_\tau^q.$$

Using 3.1 and Gronwall completes the proof. □

The first step in our integration by parts is

(3.4) PROPOSITION. *Suppose* g *is a bounded differentiable function whose gradient* Dg *is also bounded. Suppose for each* $s \le \tau$, h_s *is a functional satisfying* (2.4) *with* $c_{2.12}$ *and* $c_{2.13}$ *independent of* s, *such that* $h_s(X)$ *is right continuous in* s, a.s. *Then*

$$E\int_0^\tau Dg(X_s)^* G_s h_s(X) e_k ds = E\int_0^\tau g(X_s) h_s(X) M_s ds - E\int_0^\tau g(X_s)[h_s'(X),G] e_k ds,$$

where $M_t = \int_0^t \sigma_s^{-1} e_k dW_s$.

PROOF. Multiplying both sides of (2.6) by M_τ and taking expectations, we get

$$(3.5) \qquad EL(X)M_\tau = E\int_0^\tau E([L',\psi_\cdot^t]|\mathcal{F}_t)\sigma_t\sigma_t^{-1} e_k dt = E\int_0^\tau [L',\psi_\cdot^t] e_k dt.$$

Now let $L(X) = \int_0^\tau g(X_s)h_s(X)ds$. $[L',Y]$ is easily seen to be

$$\int_0^\tau Dg(X_s)^* Y_s h_s(X) ds + \int_0^\tau g(X_s)[h_s'(X),Y]ds.$$

Substituting in (3.5),

$$EL(X)M_\tau = E\int_0^\tau\int_0^\tau Dg(X_s)^*\psi_s^t h_s(X) ds e_k dt + E\int_0^\tau\int_0^\tau g(X_s)[h_s'(X),\psi_\cdot^t] ds e_k dt$$

$$= E\int_0^\tau Dg(X_s)^* G_s h_s(X) e_k ds + E\int_0^\tau g(X_s)[h_s'(X),G] e_k ds,$$

using Fubini and recalling that $\psi_s^t = 0$ if $s < t$, hence $\int_0^\tau \psi_s^t dt = G_s$.

Finally, let $L_t = \int_0^t g(X_s)h_s(X)ds$. L_t is a process of bounded variation. Integrating by parts ([8, p. 285]),

$$EL_\tau M_\tau = E\int_0^\tau L_s dM_s + E\int_0^\tau M_s dL_s + E\langle M,L\rangle_\tau = E\int_0^\tau g(X_s)h_s(X)M_s ds. \qquad \square$$

We next show that we can weaken the assumptions on h.

(3.6) PROPOSITION. *Suppose for each* n, $L_n(X) = \int_0^\tau g(X_s)h_n(s,X)ds$, *where*

(i) *for each* n, $h_n(s)$ *is a functional satisfying* (2.4) *with* $c_{2.12}$ *and* $c_{2.13}$ *independent of* s *(but not necessarily of* n*), such that* $h_n(s,X)$ *is right continuous in* s, a.s.;

(ii) *there is a constant* $c_{3.15}$ *such that* $sh_n(s,X) \le c_{3.15}$, a.s. *for all* n *for all* $s \le \tau$; *and* $sh_n(s,X)$ *converges to a functional* $sh_s(X)$, a.s. *for almost all* $s \le \tau$;

(iii) $c_{3.16}(q) = \sup\limits_{s\le\tau} \sup\limits_{n} |E\,[h_n'(s),G]|^q < \infty$.

Let $\varepsilon > 0$, $1 < p < 2$, *and* $\beta = 1 + \varepsilon - p/2$. *Then*

$$|E \int_0^\tau Dg(X_s)^* G_s h_s(X) e_k ds| \le c_{3.17}(\tau,\varepsilon,p)(E \int_0^\tau |g|^p(X_s)s^{-\beta}ds)^{1/p}.$$

PROOF. First of all, in (3.4) replace h_s by $h_n(s)$. Since $sh_n(s,X)$ converges boundedly to $sh_s(X)$ and $|s^{-1}G_s|^q = |J_s|^q$ has finite expectation,

$$E \int_0^\tau Dg(X_s)^* G_s h_n(s,X) e_k ds \to E \int_0^\tau Dg(X_s)^* G_s h_s(X) e_k ds.$$

Secondly, using Hölder and (iii),

$$E \int_0^\tau g(X_s)[h_n'(s,X),G]e_k ds \le c_{3.18}\Big(E \int_0^\tau |g|^p(X_s)ds\Big)^{1/p}$$

$$\le c_{3.18}\tau^{\beta/p}\Big(E \int_0^\tau |g|^p(X_s)s^{-\beta}ds\Big)^{1/p}.$$

Thirdly,

$$E \int_0^\tau g(X_s)h_n(s,X)M_s ds = E \int_0^\tau g(X_s)s^{-\beta/p}(sh_n(s,X))M_s s^{\beta/p-1}ds$$

$$\leq c_{3.19}\Big(E\int_0^\tau |g|^p(X_s)s^{-\beta}ds\Big)^{1/p}\Big(E\int_0^\tau |M_s|^q s^{(\beta/p-1)q}ds\Big)^{1/q},$$

using Hölder and (ii).

By Burkholder's inequality, $E|M_s|^q \leq c_{3.20}(q)s^{q/2}$. Our main result follows since $\int_0^\tau s^{q/2}s^{(\beta/p-1)q}ds \leq c_{3.20}(\varepsilon,p,\tau) < \infty$, noting that $q/2 + (\beta/p - 1)q = -1 + q\varepsilon/p > -1$. □

One would like to let $h_s(X) = G_s^{-1}$, but G_s^{-1} need not, except in the Markov case, be integrable. So we must construct a different h_s. The idea behind what follows is to stop the process X before $|G_s^{-1}|$ gets too large. Stopping at an ordinary stopping time would destroy the Fréchet differentiability of either L or σ; indeed, we stop at a randomized stopping time.

Let Ψ be a C^∞ function on $[0,\infty)$ such that $\Psi(x) = 1$ if $x \leq 2/3$, $\Psi(x) = 2x$ if $x \geq 3/4$, and $\Psi(x) \geq 1$ for all x. For any $d\times d$ matrix K, let $\Gamma(K) = I + (K-I)/\Psi(|K-I|)$.

Since $|\Gamma(K)-I| \leq 3/4$, $\Gamma(K)$ is invertible with inverse $\sum_{n=0}^{\infty}(I - \Gamma(K))^n$ and $|\Gamma(K)^{-1}| \leq 4$. Furthermore, if $|K-I| < 2/3$, $\Gamma(K) = K$.

Let $m_s = \sup_{r\leq s}|J_r - I|$. Let $\theta_0 = 1/4$, $\theta_1 = 1/2$, hence $(\theta_1 - \theta_0)^{-1} = 4$. Let $S_\theta = \inf\{t\colon m_t \geq \theta\}$. Observe that $m_s < \theta$ if and only if $s < S_\theta$, and if $s < S_\theta$, $\theta \leq \theta_1$, $\Gamma(J_s) = J_s$.

Define $h_s(X) = 4(m_s\vee\theta_1 - m_s\vee\theta_0)(\Gamma(J_s)^{-1})_{ki}s^{-1}$, k,i fixed. Note $|sh_s| \leq 16\theta_1$.

The main result of this section is

(3.7) THEOREM. *Let* $\varepsilon > 0$, $1 < p < 2$, $\beta = 1 + \varepsilon - p/2$. *Then*

$$\Big|E(\theta_1-\theta_0)^{-1}\int_{\theta_0}^{\theta_1}\int_0^{S_\theta\wedge\tau} D_i g(X_s)\,ds\,d\theta\Big| \leq c_{3.21}(p,\tau,\varepsilon)\Big(E\int_0^\tau |g|^p(X_s)s^{-\beta}ds\Big)^{1/p}.$$

The value of $c_{3.21}$ *depends on* X *and* σ *only through* $c_{2.2}$, $c_{2.3}$, $c_{2.4}$, *and* $c_{2.5}$ *and may be chosen so as to be an increasing function of* τ.

PROOF. Let

$$\begin{aligned} I_k &= \int_0^\tau Dg(X_s)^* G_s h_s(X) e_k ds \\ &= 4 \int_0^\tau Dg(X_s)^* G_s s^{-1} \Gamma(J_s)^{-1}_{ki} \Big(\int_{\theta_0}^{\theta_1} 1_{(\theta > m_s)} d\theta\Big) e_k ds \\ &= 4 \int_{\theta_0}^{\theta_1} \int_0^\tau Dg(X_s)^* G_s s^{-1} \Gamma(J_s)^{-1}_{ki}\, 1_{(s < S_\theta)} ds\, e_k d\theta \\ &= 4 \int_{\theta_0}^{\theta_1} \int_0^{\tau \wedge S_\theta} Dg(X_s)^* G_s (G_s^{-1})_{ki}\, e_k ds\, d\theta \\ &= (\theta_1 - \theta_0)^{-1} \int_{\theta_0}^{\theta_1} \int_0^{\tau \wedge S_\theta} (G_s^{-1})_{ki} \sum_{j=1}^d (G_s)_{jk} D_j g(X_s) ds\, d\theta, \end{aligned}$$

using the fact that $\Gamma(J_s) = J_s$ if $s < S_{\theta_1}$. Here $D_j g(X_s)$ is the jth coordinate of $Dg(X_s)$. If we sum over k, $|E \sum_{k=1}^d I_k|$ will be the desired left-hand side, since

$$\sum_j \sum_k (G_s)_{jk} (G_s^{-1})_{ki} D_j g(X_s) = D_i g(X_s).$$

It only remains to construct the appropriate sequence h_n and to apply (3.6). For fixed n, let $\Lambda_{n,\delta}$: $(\mathbb{R}^{d\times d})^n \to \mathbb{R}$ be continuously differentiable such that for all δ, $(y_1,\ldots,y_n)$, and $(z_1,\ldots,z_n)$,

$$|\Lambda_{n,\delta}(y_1,\ldots,y_n) - \Lambda_{n,\delta}(z_1,\ldots,z_n)| \le 2 \max(|y_1 - z_1|,\ldots,|y_n - z_n|)$$

and such that

$$\Lambda_{n,\delta}(y_1,\ldots,y_n) \to \max(|y_1|,\ldots,|y_n|) \quad \text{as } \delta \to 0$$

uniformly on compact sets. Let

$$m_s^n = \Lambda_{n,\delta_n}(J_{(\tau/n)\wedge s} - I, J_{(2\tau/n)\wedge s} - I, \ldots, J_{(n\tau/n)\wedge s} - I),$$

where $\delta_n \to 0$ fast enough so that for almost all s, $m_s^n \to m_s$, a.s. Let $h_n(s,X) = (m_s^n \wedge \theta_1 - m_s^n \wedge \theta_0)(\Gamma(J_s)^{-1})_{ki}(s^{-1} \wedge n)$.

Clearly (ii) of (3.6) is satisfied. Since J_s is P-Fréchet differentiable, so is m_s^n. $\| m_s^{n\prime}\|_V \le c_{3.22} \sup_{s\le\tau} |\, \| J_s'\|_V|$. Since Ψ is smooth, $\Gamma(J_s)$ is P-Fréchet differentiable, $(\Gamma(J_s)^{-1})' = \sum_{n=0}^{\infty} n(I - \Gamma(J_s))^{n-1} \cdot (\Gamma(J_s))'$. Hence $|\, \|(\Gamma(J_s)^{-1})'\|_V| \le c_{3.23} |\, \| J_s'\|_V|$. Thus (3.6i) is satisfied.

Finally,

$$\begin{aligned} E|[h_n'(s,X), G]|^q &\le E(\sup_{s\le\tau} |s\|h_n'(s,X)\|_V| \sup_{s\le\tau}|J_s|)^q \\ &\le (E \sup_{s\le\tau}|s\| h_n'(s,X)\|_V|^{2q})^{\frac{1}{2}}(E \sup_{s\le\tau}|J_s|^{2q})^{\frac{1}{2}} \\ &\le c_{3.24} < \infty \end{aligned}$$

by (3.1) and (3.3).

The assertion about $c_{3.21}$ follows by showing the corresponding assertions for $c_{3.1}, c_{3.2}, \ldots,$ noting that $|\sigma_t^{-1}| \le c_{2.3}^{-1}$. □

4.0 Densities of potentials

Throughout this section we assume σ satisfies A1 and that R1,2,3 hold. We begin by proving an elementary lemma that will be needed to handle some technical points later on. This lemma is an immediate corollary of Krylov's results on the existence of densities, but nothing so powerful is needed.

(4.1) *LEMMA.* (i) *For all* x, $E\int_0^\tau 1_{B(\varepsilon,x)}(X_s)ds \to 0$ *as* $\varepsilon \to 0$;

(ii) *For all* $r > 0$ *and all* x,

$$E\int_0^\tau 1_{[B(r+\varepsilon,x)-B(r,x)]}(X_s)ds \to 0 \quad \text{as} \quad \varepsilon \to 0.$$

PROOF. Just for the duration of this proof, let us assume without loss of generality that $X_0 = x_0 = 0$. Let X_t^i be the ith coordinate of X_t.

(i) Let $Y_t = X_t^1$. Y_t is a martingale whose diffusion coefficient $d\langle Y,Y\rangle_t/dt = \sigma_{11} > c_{4.1} > 0$. The quantity in question is less than or equal to $E\int_0^\tau 1_{[y-\varepsilon,y+\varepsilon]}(Y_s)ds$, where y is the first coordinate of x.

Let f be a function such that $f(0) = f'(0) = 0$, f'' exists and is continuous, and $1_{[y-\varepsilon,y+\varepsilon]} \le f'' \le 1_{[y-2\varepsilon,y+2\varepsilon]}$. Ito's lemma gives

$$\tfrac{1}{2}c_{4.1}E\int_0^\tau 1_{[y-\varepsilon,y+\varepsilon]}(Y_s)ds \le \tfrac{1}{2}E\int_0^\tau f''(Y_s)d\langle Y,Y\rangle_s = Ef(Y_\tau) - Ef(Y_0).$$

Since $\|f'\| \le 4\varepsilon$, $Ef(Y_\tau) - Ef(Y_0) \le \|f'\|E|Y_\tau| \to 0$ as $\varepsilon \to 0$.

(ii) Let $Y_t = |X_t - x|^2$. Using Ito's lemma, we get that Y_t is a semimartingale with drift $\sum_i 2\langle X^i - x_i, X^i - x_i\rangle_t \le c_{4.2}t$ and

$$\langle Y,Y\rangle_t = \sum_{i,j}\int_0^\tau (X_s^i - x_i)(X_s^j - x_j)\sigma_{ij}ds$$

$$\ge c_{4.3}\int_0^\tau |X_s - x|^2 ds = c_{4.3}\int_0^\tau Y_s ds,$$

using the strict ellipticity of σ. (Here x_i is the ith coordinate of x.)

Let f be a function such that $f(0) = f'(0) = 0$, f'' exists and is continuous, and $1_{[r-\varepsilon,r+\varepsilon]} \le f'' \le 1_{[r-2\varepsilon,r+2\varepsilon]}$. Applying Ito's lemma to $f(Y_t)$,

$$Ef(Y_\tau) - Ef(Y_0) \ge -E\int_0^\tau f'(Y_s)c_{4.2}ds + \tfrac{1}{2}c_{4.3}E\int_0^\tau 1_{[r-\varepsilon,r+\varepsilon]}(Y_s)(r-\varepsilon)ds.$$

Since $\|f'\| \le 4\varepsilon$ and $E|Y_\tau| < \infty$, we get the result by letting $\varepsilon \to 0$. □

Let $p_t(x,y)$ be the transition density (with respect to Lebesgue measure) for a Bessel process of index ν (recall that such a process Z_t^ν satisfies $dZ_t^\nu = dW_t + (\nu-1)/(2Z_t)dt$). We need an estimate for $U_{\tau,\beta,\nu}(x,y) = \int_0^\tau t^{-\beta} p_t(x,y)dt$, $0 < \beta < 1/2$, $\nu > 2$.

(4.2) *PROPOSITION. Let* $0 \le \lambda \le 1$, $0 < \beta < 1/2$, $\nu > 2$. *Then*

$$U_{\tau,\beta,\nu}(x,y) \le c_{4.4}(\nu,\tau,\beta)[y^{\nu-1-2(1-\lambda)\beta} 1_{(y\le 1)} + y1_{(y>1)}][1 + x^{2-\nu-2\lambda\beta}].$$

PROOF. We have that

$$p_t(x,y) = t^{-1} e^{-(x^2+y^2/2t)} (xy)^{1-\nu/2}\, I_{\nu/2-1}(xy/t) y^{\nu-1} dt, \tag{4.3}$$

where $I_p(u)$ is the modified Bessel function [5, p. 225].

It is well known that

$$I_p(u) \sim c_{4.5}(p)u^p, \ u \to 0; \qquad I_p(u) \sim c_{4.6}(p)e^u/u^{\frac{1}{2}}, \quad u \to \infty.$$

Thus, there exists $c_{4.7}(p)$ such that if $u \le 1$, $I_p(u) \le c_{4.7}(p)u^p$, while if $u \ge 1$, $I_p(u) \le c_{4.7}(p)e^u/u^{\frac{1}{2}}$. Applying this to (4.3), we get

$$\begin{aligned} U_{\tau,\beta,\nu}(x,y) &\le c_{4.7}(\nu/2-1)y^{\nu-1} \int_{xy\wedge\tau}^{\tau} t^{-\nu/2-\beta} e^{-(x^2+y^2)/2t} dt \\ &\quad + c_{4.7}(\nu/2-1)(y/x)^{(\nu-1)/2} \int_0^{xy\wedge\tau} t^{-\frac{1}{2}-\beta} e^{-(x-y)^2/2t} dt \\ &= J_1 + J_2. \end{aligned}$$

$$\begin{aligned} J_1 &\le c_{4.7}(\nu/2-1)y^{\nu-1} \int_0^\infty t^{-\nu/2-\beta} e^{-(x^2+y^2)/2t} dt \\ &\le c_{4.8}(\nu,\beta)y^{\nu-1}(x^2+y^2)^{(2-\nu-2\beta)/2}. \end{aligned}$$

If $y \geq 1$, this last expression is

$$\leq c_{4.8}(\nu,\beta) y^{\nu-1} y^{2-\nu-2\beta} \leq c_{4.8}(\nu,\beta) y.$$

If $y \leq 1$,

$$J_1 \leq c_{4.8}(\nu,\beta) y^{\nu-1-2(1-\lambda)\beta} x^{2-\nu-2\lambda\beta}, \quad 0 \leq \lambda \leq 1.$$

To investigate J_2, note that

$$e^{-(x-y)^2/2t} \leq e^{-(x-y)^2/2xy} = e^2 e^{-\frac{1}{2}(x/y+y/x)},$$

since $t \leq xy$. Note also that

$$e^{-\frac{1}{2}(x/y+y/x)}(x/y)^p \leq c_{4.9}(\nu) < \infty \quad \text{for } 0 < x,y \text{ and } -\nu-3 \leq p \leq \nu+3.$$

When $y \geq 1$,

$$J_2 \leq c_{4.10}(\nu) \int_0^{xy\wedge\tau} t^{-\frac{1}{2}-\beta} dt = c_{4.11}(\nu,\tau,\beta).$$

When $y \leq 1$,

$$J_2 \leq c_{4.12}(\nu) y^{\nu-1} x^{2-\nu} (xy)^{-\frac{1}{2}} (x/y)^{(1-2\lambda)\beta}$$

$$\cdot \int_0^{xy\wedge\tau} t^{-\frac{1}{2}-\beta} e^{-(x-y)^2/2t} (x/y)^{(\nu-2)/2-(1-2\lambda)\beta} dt$$

$$\leq c_{4.13}(\nu,\beta) y^{\nu-1} x^{2-\nu} (xy)^{-\frac{1}{2}} (x/y)^{(1-2\lambda)\beta} \int_0^{xy} t^{-\frac{1}{2}-\beta} dt$$

$$\leq c_{4.14}(\nu,\beta) y^{\nu-1-2(1-\lambda)\beta} x^{2-\nu-2\lambda\beta}.$$

Summing, we get our result. □

We next derive a connection between the right-hand side of (3.7) and Bessel processes. Recall $X_0 = x_0$, a.s.

(4.4) *THEOREM. Let* $w = (w_1,\dots,w_d)$ *be fixed and suppose* $w \neq x_0$. *Suppose* $\hat{g}(y) = \sup_{|z-w|=y} |g(z)|$ *is nonincreasing in* y. *Suppose* $0 < \beta < 1/2$ *and*

$$2 < \nu < 1 + (1 + c_{2.11})^{-2}[(d - 1) - (d + 1)c_{2.11}].$$

Then,

$$E \int_0^\tau |g|^p(X_t)t^{-\beta}dt \le c_{4.15} \int_0^\infty |\hat{g}|^p(y)U_{2\tau,\beta,\nu}(|x_0 - w|,y)dy.$$

PROOF. Let $Y_t = |X_t - w|$. Using Ito's lemma for $t \le \inf\{t:|X_t - w| < 1/n\}$, we get that Y_t satisfies $Y_0 = |x_0 - w|$ and

$$dY_t = \sum_{i,j} \frac{X_t^i - w_i}{|X_t - w|} \sigma_{ij} dW_t^j + \frac{1}{2} \sum_{i,j} \frac{1_{(i=j)}|X_t - w|^2 - (X_t^i - w_i)(X_t^j - w_j)}{|X_t - w|^3} \sigma_{ij} dt.$$

Thus Y_t is a semimartingale with diffusion coefficient

$$a_t = d\langle Y,Y\rangle/dt = (X_t - w)^*\sigma\sigma^*(X_t - w)/|X_t - w|^2$$

and drift coefficient $b_t/(2Y_t)$, where

$$b_t = \text{trace } \sigma - ((X_t - w)^*\sigma(X_t - w)/|X_t - w|^2).$$

Using R1, $b_t/a_t > \nu - 1$.

We now time change Y_t. Let $C_t = \langle Y,Y\rangle_t^{-1}$. Note $\frac{1}{2} \le dC_t/dt \le 2$. Let $Z_t = Y_{C_t}$. Z_t is a semimartingale adapted to $G_t = \sigma(X_{C_s};\ s \le t)$.

Checking that

$$Z_t - \frac{1}{2}\int_0^t b_{C_s}/(a_{C_s}Z_s)ds = Y_{C_t} - \frac{1}{2}\int_0^{C_t} b_r/Y_r dr$$

is a G_t-martingale with $\langle Z,Z\rangle_t = t$, we see that Z_t is equal to a one-dimensional Brownian motion $\hat{W}_t$ plus $\frac{1}{2}\int_0^t b_{C_s}/(a_{C_s}Z_s)ds$. By a comparison theorem for stochastic differential equations (for example, see [5, p. 352]), we conclude that $Z_t \geq Z_t^\nu$ for all $t \leq \inf\{t:|X_t-w| < 1/n\}$, where Z_t^ν is a Bessel process of index ν defined in terms of $\hat{W}_t$. Since $\nu > 2$, Z_t^ν never hits 0, hence Z_t never does either, hence $Z_t \geq Z_t^\nu$ for all t.

Finally, using $\frac{1}{2} \leq dC_s/ds \leq 2$ and $|\hat{g}|$ nonincreasing,

$$\begin{aligned} E\int_0^\tau |g|^p(X_t)t^{-\beta}dt &\leq E\int_0^\tau |\hat{g}|^p(Y_t)t^{-\beta}dt \\ &= E\int_0^{\langle Y,Y\rangle_\tau} |\hat{g}|^p(Z_t)C_t^{-\beta}dC_t \\ &\leq c_{4.16}(\beta)E\int_0^{2\tau}|\hat{g}|^p(Z_t)t^{-\beta}dt \\ &\leq c_{4.16}(\beta)E\int_0^{2\tau}|\hat{g}|^p(Z_t^\nu)t^{-\beta}dt \\ &\leq c_{4.16}(\beta)\int_0^\infty|\hat{g}|^p(y)U_{2\tau,\beta,\nu}(|x_0-w|,y)dy. \qquad \square \end{aligned}$$

We come now to the main result of this section. Let S_θ, θ_0, θ_1 be as in Section 3. Define

$$\lambda(F) = E(\theta_1-\theta_0)^{-1}\int_{\theta_0}^{\theta_1}\int_0^{\tau\wedge S_\tau} 1_F(X_s)ds\,d\tau.$$

(4.5) *THEOREM. Suppose* $1 < p < 2$, $\varepsilon < 0$, $0 < \lambda < 1$, $d-1 < \nu < d$, *and* ν *satisfies the hypothesis of* (4.4). *Let*

$$\beta = 1+\varepsilon-p/2, \qquad \gamma = (1-d)p+\nu-2(1-\lambda)\beta,$$

and suppose $\beta < \frac{1}{2}$, $\gamma > 0$.

Then, $\lambda(F)$ *has a density* $\rho(z)$ *with respect to Lebesgue measure which satisfies*

(i) $|\rho(z)| \le c_{4.17}(1 + |z - x_0|^{(2-\nu-2\lambda\beta)/p})$;

and

(ii) $|\rho(z) - \rho(z')| \le c_{4.18}\zeta^{-c_{4.19}}|z - z'|^{\gamma/(p+\lambda)}$,

where $\zeta = \min(|z - x_0|, |z' - x_0|, 1)$.

COMMENT. In (5.2) we show how to select p, ε, λ, and ν to satisfy the constraints we have put on them.

PROOF. Let $\rho_r(z) = r^{-d}\lambda(B(r,z))$. The first step is to show that $\rho_r(z)$ converges as $r \to 0$. Fix z. Let

$$\begin{aligned} g_{ir}(y) &= (y_i - z_i)/|y - z|^d \quad \text{if } |y - z| > r, \\ &= (y_i - z_i)/r^d \quad \text{if } |y - z| \le r. \end{aligned}$$

Note that $|g_{ir}(y)| \le |y - z|^{1-d}$ and that $\sum_i D_i g_{ir}(y) = dr^{-d}$ if $|y - z| < r$, 0 if $|y - z| > r$.

(3.7) is not valid for g_{ir} since g_{ir} is not differentiable at $|y - z| = r$. However, g_{ir} is uniformly Lipschitz; an easy approximation argument together with (4.1ii) shows that (3.7) may be applied to g_{ir}. We do so, and summing over i, we get

$$\begin{aligned} (4.6) \qquad |d\rho_r(z)| &\le c_{4.20}(p,\tau,\varepsilon)\sum_{i=1}^{d}\Big(E\int_0^\tau |g_{ir}|^p(X_t)t^{-\beta}dt\Big)^{1/p} \\ &\le c_{4.21}\Big(E\int_0^\tau |X_t - z|^{(1-d)p}\, t^{-\beta}dt\Big)^{1/p} \\ &\le c_{4.22}\Big(\int_0^\infty y^{(1-d)p}\, U_{2\tau,\beta,\nu}(|x_0 - z|, y)dy\Big)^{1/p} \\ &\le c_{4.23}\big(1 + |x_0 - z|^{2-\nu-2\lambda\beta}\big)^{1/p}, \end{aligned}$$

using (4.2), R3, and the hypothesis that $\gamma > 0$.

Now applying (3.7) to $g_{ir} - g_{is}$ and summing over i, we get

$$(4.7)\quad d|\rho_r(z) - \rho_s(z)| \le c_{4.24}\sum_{i=1}^{d}\Big(E\int_0^{\tau}|g_{ir} - g_{is}|^p(X_t)t^{-\beta}dt\Big)^{1/p}$$

$$\le c_{4.25}\Big(E\int_0^{\tau}|X_t - z|^{p(1-d)}1_{[0,r\vee s]}(|X_t - z|)t^{-\beta}dt\Big)^{1/p}$$

$$\le c_{4.26}\Big(\int_0^{r\vee s} y^{(1-d)p}\, U_{2\tau,\beta,\nu}(|x_0 - z|,y)dy\Big)^{1/p}$$

$$\le c_{4.27}\Big(\big(|x_0 - z|^{2-\nu-2\lambda\beta} + 1\big)(r\vee s)^{\gamma}\Big)^{1/p}.$$

Here we use the fact that $g_{ir} - g_{is}$ is 0 if $|y - z| \ge r \vee s$.

Thus, $\rho_r(z)$ converges, say to $\rho(z)$. Taking the limit in (4.6) gives (i), while taking the limit in (4.7) gives

$$(4.8)\qquad |\rho_r(z) - \rho(z)| \le c_{4.28}\big(|x_0 - z|^{(2-\nu-2\lambda\beta)/p} + 1\big)r^{\gamma/p}.$$

The next step is to show $\rho(z)$ is continuous in z.

$$\lambda(B(r,z') - B(r,z)) \le \lambda(B(r + |z - z'|,z) - B(r,z)) \to 0$$

as $|z - z'| \to 0$ by (4.1ii). But

$$|\rho_r(z) - \rho_r(z')| \le r^{-d}\lambda(B(r,z)\,\Delta B(r,z')),$$

and so $\rho_r(z)$ is continuous in z. (Δ denotes symmetric difference.) By (4.8), $\rho(z)$ is the uniform limit of $\rho_r(z)$ in regions bounded away from x_0, and hence ρ is continuous in such a region.

It follows (cf. [9, Ch. 8]) that ρ is a density for λ in any region bounded away from x_0. Since (4.1i) shows that λ assigns no mass to $\{x_0\}$, ρ is a density for λ over all of $\mathbb{R}^d$.

We now establish (ii). Let $\zeta = \min(|x_0 - z|, |x_0 - z'|, 1)$, $\delta = |z - z'|$, and suppose $r,\delta < \zeta/6$. Since $B(r,z') - B(r,z) \subseteq B(r+\delta,z) - B(r,z)$, it follows that the Lebesgue measure of $B(r,z')\,\Delta B(r,z)$ is $\le c_{4.29}\,\delta r^{d-1}$ as long as $\delta < r/2$.

Then
$$|\rho_r(z) - \rho_r(z')| \le r^{-d}\lambda(B(r,z)\,\Delta B(r,z'))$$
$$= r^{-d}\int_{B(r,z)\Delta B(r,z')} \rho(w)dw$$
$$\le c_{4.30} r^{-d}\delta r^{d-1}\,\zeta^{(2-\nu-2\lambda\beta)/p}$$
$$\le c_{4.31} r^{-1}\,\delta\zeta^{-c_{4.32}},$$

since $\rho(w)$ is bounded as long as $|w - x_0| > \zeta/3$.

Together with (4.8),

$$|\rho(z) - \rho(z')| \le c_{4.33}\zeta^{-c_{4.32}}(\delta r^{-1} + r^{\gamma/p}).$$

Letting $r = \delta^{p/(p+\gamma)}$, $\delta = r^{(p+\gamma)/p} < r/2$ if δ is small enough, and then

$$|\rho(z) - \rho(z')| \le c_{4.34}\zeta^{-c_{4.32}}\,\zeta^{\gamma/(p+\gamma)}.$$

Since $\gamma/(p+\gamma) > 0$, this proves (ii). □

5.0 Potentials of occupation times

In this section we will assume that R1,2,3 hold. We will also assume that σ satisfies A1 until (5.4), after which we will allow σ to satisfy A2 instead.

We state some elementary results. Recall that ϕ_b is the density of a normal random variable with mean 0 and variance b.

(5.1) *PROPOSITION. Let* λ, μ *be measures satisfying* M1:

$$\lambda(\mathbb{R}^d),\ \mu(\mathbb{R}^d) \le c_{2.6}$$

$$\lambda(B(r,x)),\mu(B(r,x)) \le c_{2.7} r^{d-2+c_{2.8}} \quad \textit{for all } x,r.$$

Then, (i) $\mu * \phi_b(dx)$ *has density* $\int \phi_b(x-y)\mu(dy)$, *which is bounded and uniformly continuous in* x;

(ii) $\mu * \phi_b$, $\lambda * \phi_b$ *satisfy* M1 *with the same constants* $c_{2.6}$, $c_{2.7}$, *and* $c_{2.8}$;

(iii) $\|\mu * \phi_a - \mu * \phi_b\|_{W-\alpha} \le c_{5.1}|b-a|^{\alpha}$, *where* $c_{5.1}$ *depends only on* α, $c_{2.6}$, $c_{2.7}$, *and* $c_{2.8}$;

(iv) *If* $\|\mu - \lambda\|_{W-\alpha} \le c_{5.2}$, *then* $\|\mu * \phi_b - \lambda * \phi_b\|_{W-\alpha} \le c_{5.2}$;

Let $p,\varepsilon,\nu,\beta,\gamma,\rho(z)$ *be as in* (4.3). *Suppose*

$$(\nu - 2 + 2\lambda\beta)/p < d - 2 + c_{2.8}.$$

Then

(v) $\int \rho(z)\lambda(dz) \le c_{5.3}$, *where the value of* $c_{5.3}$ *depends on* λ *only through* $c_{2.6}$, $c_{2.7}$, *and* $c_{2.8}$;

(vi) $\int \rho(z)(\lambda-\mu)(dz) \le c_{5.4}\|\lambda-\mu\|^{c_{3.5}}_{W-\gamma/(p+\gamma)}$, *where the value of* $c_{5.4}$ *depends on* λ *and* μ *only through* $c_{2.6}$, $c_{2.7}$, *and* $c_{2.8}$.

PROOF. Very similar results are proved in section 3 of [1]. □

It is not clear that $\nu,\beta,\lambda,\varepsilon,p$ may be selected to satisfy all the constraints we have placed on them. We pause to verify that we can do so.

(5.2) *PROPOSITION. If* R1 *holds,* ε, p, ν, *and* λ *may be selected so that*

(i) $\beta < 1/2$;

(ii) $\gamma > 0$;

(iii) $(\nu - 2 + 2\lambda\beta)/p < d - 2 + c_{2.8}$,

where β and γ were defined in (4.5).

PROOF. Under R1, we may select ν in (4.4) so that

$$d - 1/3(c_{2.8} \wedge 1) < \nu < d + 1/3(c_{2.8} \wedge 1)$$

by taking ν less than but close to $1 + (1 + c_{2.11})^{-2}[(d-1) - (d+1)c_{2.11}]$.

Let $\lambda = \frac{1}{2}(c_{2.8} \wedge 1)$. Take ε small enough so that $\nu + \lambda - 2\varepsilon(1-\lambda) > d$, choose p close to 1 so that

$$1 < p < (\nu - (1-\lambda)(2+2\varepsilon))/(d + \lambda - 2),$$

and then, if necessary, choose ε even smaller so that $1 + \varepsilon - p/2 < 1/2$. (i), (ii), and (iii) now follow. □

If we apply (5.1v,vi) to $\mu * \phi_b, \mu * \phi_a$, we get

$$\left| E(\theta_1 - \theta_0)^{-1} \int_{\theta_0}^{\theta_1} \int_0^{S_\theta \wedge \tau} \mu * \phi_b(X_s) ds \, d\theta \right| \le c_{5.3} \quad \text{and}$$

$$\left| E(\theta_1 - \theta_0)^{-1} \int_{\theta_0}^{\theta_1} \int_0^{S_\theta \wedge \tau} (\mu * \phi_b - \mu * \phi_a)(X_s) ds \, d\theta \right| \le c_{5.6} |b - a|^{c_{5.7}}.$$

Here $\mu * \phi_b(x)$ is the density of $\mu * \phi_b$ evaluated at x.

Let $\tilde{\theta}_0$ be a random variable that is uniform on $[\theta_0, \theta_1]$ and independent of W_t. In the usual way, we enlarge the probability space and redefine the probability P and the sigma fields F_t to ensure that this is possible. If $S^0 = \inf\{t: m_t \ge \tilde{\theta}_0\}$, what we have proved is that

$$\left| E \int_0^{S^0 \wedge \tau} \mu * \phi_b(X_s) ds \right| \le c_{5.3}$$

$$\left| E \int_0^{S_0 \wedge \tau} (\mu * \phi_b - \mu * \phi_a)(X_s) ds \right| \le c_{5.6} |b - a|^{c_{5.7}}.$$

We now attempt to eliminate the S^0 from the above two expressions.

(5.3) *PROPOSITION.* (i) $|E\int_0^\tau \mu * \phi_b(X_s)ds| \le c_{5.8}$

(ii) $|E\int_0^\tau (\mu * \phi_b - \mu * \phi_a)(X_s)ds| \le c_{5.9}|b - a|^{c_{5.7}}$, *where* $c_{5.7}$, $c_{5.8}$, *and* $c_{5.9}$ *depend on* X, σ, *and* μ *only through* $c_{2.2} - c_{2.8}$, $c_{2.11}$.

PROOF. We prove (i), (ii) being similar. Let $Q_\omega^0(\cdot)$ be a regular conditional probability distribution (r.c.p.d) for $E(\cdot|F_{S^0})$. Since X_{S^0} and S^0 are F_{S^0}-measurable, we claim that for each ω', $\hat{X}_t(\omega) = X_{S^0+t}(\omega)$ is a d-dimensional martingale under $Q_{\omega'}^0$, and the diffusion coefficients $\hat{\sigma}$ will still satisfy A1 and R1. In fact, under $Q_{\omega'}^0$, $\hat{\sigma}_t(\omega) = \sigma_{t+S^0(\omega')}(Y)$ where $Y_s(\omega) = X(\omega')$ if $s \le S^0(\omega')$, $Y_s(\omega) = \hat{X}_{s-S^0(\omega')}(\omega)$ if $s > S^0(\omega')$.

Now let $S^1 = \inf\{t: \hat{m}_t \ge \tilde{\theta}_1\}$, where $\tilde{\theta}_1$ is an r.v. that is uniform on $[\theta_0,\theta_1]$, and, under Q_ω^0, independent of $\tilde{\theta}_0, W_t$; $\hat{m}_t$ is defined in a manner analogous to m_t. We then get, as before,

$$|Q_\omega^0 \int_0^{S^1\wedge(\tau-S^0)} \mu * \phi_b(X_s)ds| \le c_{5.3}.$$

Then

$$|E\int_0^{(S^0+S^1)\wedge\tau} \mu * \phi_b(X_s)ds| \le |E\int_0^{S^0\wedge\tau} \mu * \phi_b(X_s)ds|$$

$$+ |E\big(|Q_\omega^0 \int_0^{S^1\wedge(\tau-S^0)} \mu * \phi_b(\hat{X}_s)ds|;\ S^0 \le \tau\big)|$$

$$\le c_{5.3} + c_{5.3}\, P(S^0 \le \tau).$$

Repeat, letting Q_ω^1 be an r.c.p.d. for $E(\cdot|F_{S^0+S^1})$, etc. By induction, we get

$$|E\int_0^{(S^0+\cdots+S^n)\wedge\tau} \mu * \phi_b(X_s)ds| \le c_{5.3}(1 +\cdots+ P(S^0 +\cdots+ S^n \le \tau)).$$

To complete the proof, it suffices to show that

(a) $S^0 +\cdots+ S^n \to \infty$, a.s., and

(b) $\sum_{n=0}^{\infty} P(S^0 +\cdots+ S^n \le \tau) \le c_{5.10} < \infty$.

By (3.1iv),

$$Ee^{-(S^0+S^1)} = Ee^{-S^0}(Q^0_\omega e^{-S^1}) \le c_{3.4} Ee^{-S^0} \le c_{3.4}{}^2.$$

By induction, $Ee^{-(S^0+\cdots+S^n)} \le c_{3.4}{}^n \to 0$, which proves (a).

$$P(S^0 +\cdots+ S^n \le \tau) \le Ee^{-(S^0+\cdots+S^n)}/e^{-\tau} \le e^\tau c_{3.4}{}^n;$$

using Chebyshev, and summing over n gives (b).

(5.4) *PROPOSITION.* (5.3) *holds if* σ *only satisfies* A2.

PROOF. Take a sequence of σ_n's converging to σ as in A2. Let X^n be the solution to $X^n_0 = x_0$, $dX^n_t = \sigma^n_t(X^n)dW_t$, and let P^n be the law induced on $C[0,\infty)$ by X^n. Let P^0 be the law induced on $C[0,\infty)$ by X. Since X is the unique pathwise solution of a stochastic differential equation, P^0 satisfies a uniqueness in law property [10], and it is not hard to show that P^n converges weakly to P^0.

Since $\mu * \phi_b$ is continuous, $L(f) = \int_0^\tau \mu * \phi_b(f(s))ds$ is a continuous functional on $C[0,\tau]$. But

$$\left|E\int_0^\tau \mu * \phi_b(X_s)ds\right| = |P^0L(X)| = \lim_{n\to\infty}|P^nL(X)|$$
$$= \lim_{n\to\infty}\left|E\int_0^\tau \mu * \phi_b(X^n_s)ds\right| \le c_{5.8},$$

independent of n.

The proof of (5.3ii) is similar. □

We also need (5.4) for conditional probabilities.

(5.5) *THEOREM. If* $t \le \tau$,

(i) $\left|E\left(\int_t^\tau \mu * \phi_b(X_s)ds \middle| F_t\right)\right| \le c_{5.8}$, a.s.;

(ii) $\left|E\left(\int_t^\tau (\mu * \phi_b - \mu * \phi_a)(X_s)ds \middle| F_t\right)\right| \le c_{5.9}|b - a|^{c_{5\ 7}}$, a.s.

PROOF. Let Q_ω be an r.c.p.d. for $E(\cdot|F_t)$. Letting $\hat{X}_s = X_{t+s} = X_t + \int_t^{t+s} \sigma(X_r)dW_r$, one checks as above that $(\hat{X}_s, Q_\omega)$ satisfies A2 and R1,2,3, and hence

(5.6) $$\left|Q_\omega \int_0^{t-\tau} \mu * \phi_b(\hat{X}_s)ds\right| \le c_{5.11}(\tau - t).$$

$c_{5.11}$ may be chosen to be an increasing function of $\tau - t$, and hence is $\le c_{5.12}(\tau)$, independent of t. But this is just what we needed.

(ii) is similar. □

Our final step is the following.

(5.7) *THEOREM. Suppose* A *and* B *are two increasing processes,* C = A - B. *Suppose* $|E(A_\tau - A_t|F_t) + E(B_\tau - B_t|F_t)| \le N$ *for all* $t \le \tau$ *and that* $|U_t| \le \varepsilon < 1$ *for all* $t \le \tau$, *where* $U_t = E(C_\tau - C_t|F_t)$. *Then*

(i) $E(\sup_{t\le\tau}|C_t|)^p \le c_{5.13}(N,p)\varepsilon^{p/2}$ *and*

(ii) $E(\sup_{t\le\tau}|A_t|^p) \le c_{5.14}N^p$.

PROOF. First of all,

$$(C_\tau - C_t)^2 = 2\int_t^\tau (C_\tau - C_s)dC_s,$$

and so,

$$E((C_\tau - C_t)^2|F_t) = 2E\Big(\int_t^\tau E(C_\tau - C_s|F_s)dC_s|F_t\Big)$$

$$\le 2\varepsilon E\Big(\int_t^\tau d|C_s|\ |F_t\Big)$$

$$\le 2\varepsilon N.$$

Secondly, if $M_t = E(C_\tau|F_t)$, $U_t = M_t - C_t$, and if $t \le \tau$,

$$E(<M,M>_\tau - <M,M>_t|F_t) = E((M_\tau - M_t)^2|F_t)$$

$$\le 2U_t^2 + 2E((C_\tau - C_t)^2|F_t)$$

$$\le 2\varepsilon^2 + 4\varepsilon N \le c_{5.15}(N)\varepsilon.$$

Integrating by parts gives

$$<M,M>_\tau^n = n\int_0^\tau (<M,M>_\tau - <M,M>_t)d<M,M>_t^{n-1},$$

and

$$E<M,M>_\tau^n = nE\int_0^\tau E(<M,M>_\tau - <M,M>_t|F_t)d<M,M>_t^{n-1}$$

$$\le n\ c_{5.15}(N)\varepsilon\ E\int_0^\tau d<M,M>_t^{n-1}.$$

By induction,

$$E<M,M>_\tau^p \le c_{5.15}{}^p(N)\varepsilon^p\ p!\,.$$

Finally, since $C_t = M_t - U_t$,

$$E\sup_{t\le\tau}|C_t|^p \le c_{5.16}(p)E\sup_{t\le\tau}|M_t|^p + c_{5.16}(p)E\sup_{t\le\tau}|U_t|^p$$

$$\le c_{5.17}(p)E<M,M>_\tau^{p/2} + c_{5.16}(p)\varepsilon^p \quad \text{(Burkholder)}$$

$$\le c_{5.18}\varepsilon^{p/2}.$$

(ii) is similar. □

6.0 Proof of theorems

In this section we prove T1,2,3,4 under the assumptions that A2 and R1,2,3 hold.

PROOF of T1. Abbreviate $A_t(\mu * \phi_b)$ by A_t^b. Combining (5.5) and (5.7),

$$E \sup_{t \le \tau} |A_t^b - A_t^a|^p \le c_{6.1}(p)|b - a|^{c_{6.2}p}.$$

Take p large enough so that $c_{6.2}p > 1$. Kolmogorov's theorem then implies that there exist versions of A_t^b such that A_t^b is uniformly jointly continuous, a.s., $0 \le t \le \tau$, $0 < b \le 1$. Since the density of $\mu * \phi_b$ is uniformly continuous in b for b in a closed interval not containing 0, there is, in fact, no need to take versions. It is then immediate that there exists a process $A_t(\mu)$ that is the uniform limit of $A_t(\mu * \phi_b)$, a.s. □

If we replaced $\mu * \phi_b(y)dy$ by some other set of measures converging appropriately to μ, one would like to know that one gets the same limit process $A_t(\mu)$. This will follow from the proof of T2 below.

(6.1) *COROLLARY. Suppose* μ *satisfies* M1(ii), *but we replace* M1(i) *by* $\int_{\mathbb{R}^d} e^{-|y|^2} \mu(dy) < \infty$.

Then there exists a process $A_t(\mu)$ *such that* $\sup_{t \le \tau} |A_t(\mu * \phi_b) - A_t(\mu)| \to 0$.

PROOF. If $T = \inf\{t: |X_t| \ge M\}$, it suffices to show $A_t(\mu * \phi_b)$ converges uniformly for $t \le \tau \wedge T$, and then let $M \to \infty$. But if $\mu_0 = \mu|_{B(2M,0)}$ (the restriction of μ to $B(2M,0)$) and $\mu_1 = \mu|_{B^c(2M,0)}$, then

$$A_t(\mu * \phi_b) = A_t(\mu_0 * \phi_b) + A_t(\mu_1 * \phi_b).$$

μ_0 satisfies M1, and so the first term on the right converges.

$$\mu_1 * \phi_b(x) = \int_{B^c(2M,0)} \phi_b(x - y)\mu_1(dy) \to 0$$

uniformly as $b \to 0$ for $|x| \le M$, and so

$$A_{\tau \wedge T}(\mu_1 * \phi_b) = \int_0^{\tau\wedge T} \mu_1 * \phi_b(X_s) \to 0 \quad \text{a.s.}$$

as $b \to 0$. □

If we replace ϕ_b by an approximation to the identity that has compact support, we can define $A_t(\mu)$ for any μ locally bounded and satisfying M1(ii).

Before proving T2, we need the following technical lemma.

(6.2) *Suppose* $0 < \alpha, \beta \le 1$, μ *a signed measure*, $\|\mu\|_V \le c_{6.3}$, *and* $\|\mu\|_{W-\alpha} \le \delta$. *Then* $\|\mu\|_{W-\beta} \le c_{6.4}\, \delta^{c_{6.5}}$, $c_{6.4}$, $c_{6.5}$ *depending only on* α, β, *and* $c_{6.3}$.

PROOF. Suppose $\|f\|_{C^\beta} \le 1$.

$$|\int f\, d\mu| \le |\int (f - f * \phi_b) d\mu| + |\int f * \phi_b\, d\mu|. \tag{6.3}$$

Since $\|f\|_{C^\beta} \le 1$, $\|f - f * \phi_b\| \le c_{6.6} b^\beta$, and so the first term on the right of (6.3) is $\le c_{6.3}\, c_{6.6}\, b^\beta$. If $b \le 1$, $\|f * \phi_b\|_{C^\alpha} \le c_{6.7}\, b^{-c_{6.8}}$ and so the second term on the right of (6.3) is $\le c_{6.7} \delta b^{-c_{6.8}}$.

Summing, and then letting $b = \delta^{1/(\beta + c_{6.8})}$ gives

$$|\int f\, d\mu| \le c_{6.9}\, \delta^{\beta/(\beta + c_{6.8})}.$$ □

PROOF of T2. In view of (6.2), we may suppose that the α in the statement of T2 is equal to $\gamma/(\beta + \gamma)$, γ as in (4.5). By (5.1) and the fact that if $b < b'$, $\phi_{b'} = \phi_b * \phi_{b'-b}$,

$$\| \mu_a * \phi_b - \mu_{a'} * \phi_{b'} \|_{W-\alpha} \le c_{6.10}(|a - a'|^{c_{6.11}} + |b - b'|^{c_{6.11}}).$$

Applying (5.1v,vi), repeating the arguments of (5.3), (5.4) and (5.5), and then applying (5.7),

$$E \sup_{t \le \tau} |A_t^{b,a} - A_t^{b',a'}|^p \le c_{6.12}(p)|(a,b) - (a',b')|^{c_{6.11}p},$$

where $A_t^{b,a} = \int_0^t \mu_a * \phi_b(X_s)ds$.

If p is chosen large enough so that $c_{6.11}p > 2$, then by the two parameter version of Kolmogorov's theorem, there exist versions of $A_t^{b,a}$ that are jointly uniformly continuous, a.s., $t \le \tau$, $0 \le a \le 1$, $0 < b \le 1$. Again, it is immediate that there exist processes A^a such that A^a is the uniform limit of $A^{b,a}$, $b \to 0$, and A^a is jointly continuous in t and a. □

Note that

$$E \sup_{t \le \tau} |A_t(\mu_a) - A_t(\mu_b)|^p \le c_{6.12}|b - a|^{c_{6.11}p}$$

Using the generalization of the lemma of Garsia, Rodemich, and Rumsey [10, p. 60], one gets that there exists a $\delta > 0$ and a random variable $H(\omega) < \infty$, a.s. such that

$$\sup_{t \le \tau} |A_t(\mu_a)(\omega) - A_t(\mu_b)(\omega)| \le H(\omega)|b - a|^{\delta} \quad \text{for all } a,b.$$

We now prove T3 and T4. The proofs are those of [1]; we here mainly point out the necessary modifications.

PROOF of T3. Let T be a stopping time $\le \inf\{t: |X_t| \ge M\} \wedge \tau$. Let $K: [0,\infty) \to \mathbb{R}$ such that K has support in $[0,1]$, K is continuous, and $\int_{\mathbb{R}^d} K(|y|)dy = 1$. Let $K_a(y) = a^{-d}K(|y|/a)$, $1 > a > 0$.

Let $\lambda(s,v)$ be $d-1$ dimensional Lebesgue measure on the hyperplane $\{y: v*y = s\}$, where $s \in \mathbb{R}$, $v \in \mathbb{R}^d$, and $|v| = 1$.

First of all, $A_t(\lambda(s,v))$ is jointly continuous in t, s, v, a.s. This may be proved as in the proof of T2 and [1, section 2].

Secondly, if $a \le 1$, $A_T(\lambda(s,v) * K_a) = A_T(\lambda(s,v)|_{B(M+1,0)} * K_a)$. In particular, $A_T(\lambda(s,v) * K_a) = 0$ if $|s| \ge M+1$. Then $A_t(\lambda(s,v) * K_a) \to A_t(\lambda(s,v))$ uniformly in s, v, $t \le T$ by T2. We also get

$$E\, A_T^2(\lambda(s,v) * K_a) \le E\, A_T^2(\lambda(s,v)|_{B(M+1,0)} * K_a) \le c_{6.13},$$

$c_{6.13}$ independent of s, v, a, and therefore

$$E\, A_T(\lambda(s,v) * K_a) \to E\, A_T(\lambda(s,v)).$$

Now if $a \le 1$, let

$$f_a(x) = E\int_0^T K_a(X_s - x)ds.$$

f_a is bounded and has support in $B(M+1,0)$. Apply the Radon transform formula [1, section 4] to f_a to get

(6.4) $f_a * \phi_b(y) =$

$$(2\pi)^{-d} \int_{(|v|=1)} \int_0^\infty \int_{-\infty}^\infty e^{iq(s-v*y)} q^{d-1} e^{-bq^2/2} [\int f_a(z)\lambda(s,v)(dz)]ds\, dq\, dv.$$

Integrate both sides of (6.4) with respect to $\mu(dy)$. The left-hand side is

$$\int f_a * \phi_b(y)\mu(dy) = E\int_0^T [\int K_a(X_t - y)\mu * \phi_b(y)dy]dt$$

$$\to E\int_0^T \mu * \phi_b(X_t) \quad \text{as} \quad a \to 0$$

since $\mu * \phi_b$ is bounded and uniformly continuous, hence $K_a * (\mu * \phi_b) \to \mu * \phi_b$ uniformly.

On the right-hand side of (6.4),

$$\int f_a(z)\lambda(s,v)(dz) = E \int \int_0^T K_a(X_t - z)dt\, \lambda(s,v)(dz)$$

$$= E\, A_T(\lambda(s,v) * K_a)$$

$$\to E\, A_T(\lambda(s,v)) \quad \text{as} \quad a \to 0.$$

Using dominated convergence and Fubini, we get

$$E\, A_T(\mu * \phi_b) = E\, \hat{B}_T(\mu,b).$$

The same argument shows that

$$E(A_T(\mu * \phi_b) - A_t(\mu * \phi_b)|F_t) = E(\hat{B}_T(\mu,b) - \hat{B}_t(\mu,b)|F_t) \quad \text{on } t \leq T;$$

therefore $A_t(\mu * \phi_b) - \hat{B}_t(\mu,b)$, $t \leq T$, is a continuous martingale with paths of bounded variation, hence 0, a.s. Finally, $A_t(\mu * \phi_b) \to A_t(\mu)$, $t \leq T$ by T1.

To finish the proof, it suffices to take a sequence of such stopping times T increasing to τ. Under R1,2,3, we may use the sequence $T_M = \inf\{t: |X_t| \geq M\} \wedge \tau$. □

PROOF of T4. By a method identical to that of [1], we get

(6.5) $E(A_T(\mu * \phi_b)) =$

$$E\tfrac{1}{2}(-1)^{(d-1)/2} \int_{(|v|=1)} \int_{-\infty}^{\infty} D_{d-1} g_v(s)[A_T(\lambda(s,v))]ds\, dv + J_b,$$

where T is a stopping time $\leq \inf\{t: |X_t| \geq M\} \wedge \tau$ and J_b is a term that $\to 0$ as $b \to 0$.

$$E\, A_T^2(\mu * \phi_b) \le E\, A_\tau^2(\mu * \phi_b) \le c_{6.14}.$$

So the left-hand side of (6.5) approaches $E\, A_T(\mu)$ as $b \to 0$.

We get a similar equation for $E(A_T(\mu) - A_t(\mu)|F_t)$, $t \le T$, and arguing as in the proof of T3, we see that our result holds for $t \le T$. Then take the appropriate sequence of T's. □

One could, of course, weaken the hypothesis that $\mu(\mathbb{R}^d) < \infty$ as in (6.1).

7.0 Removal of Restrictions

In this section we show that T1,2,3, and 4 still hold when restrictions R1,2, and 3 are removed.

REMOVAL OF R1. Given a measure(s) μ (μ_a) satisfying M1, let ε be chosen so that

$$2\varepsilon < (d+1)^{-1}(c_{2.8} \wedge 1)/16.$$

Define a C^∞ function $\Gamma\colon \mathbb{R}^{d\times d} \to \mathbb{R}^{d\times d}$ such that $|\Gamma(K) - I| < 2\varepsilon$ for all K and $\Gamma(K) = K$ if $|K - I| < \varepsilon$. (Cf. construction following (3.6).)

Consider the process $\hat{X}_t$ that solves $\hat{X}_t = x_0$, $d\hat{X}_t = \Gamma(\sigma(\hat{X}))dW_t$. $\Gamma(\sigma)$ satisfies A2 and R1. Clearly $X_t = \hat{X}_t$ up to time $T_1 = \inf\{t\colon |\sigma_t - I| > \varepsilon\} \wedge \tau$. Applying T1 to the process $\hat{X}_t$, we get that $A_t(\mu * \phi_b)$ converges uniformly to a process $A_t(\mu)$, $t \le T_1$. Also, if $t \le T_1$,

$$E(|A_{T_1}(\mu * \phi_b) - A_t(\mu * \phi_b)|^p|F_t) \le E\Big(\Big|\int_t^\tau \mu * \phi_b(\hat{X}_s)ds\Big|^p\Big|F_t\Big)$$

$$\le c_{7.1}(p).$$

Let C_1 be a square root of $\sigma_{T_1}^{-1}$, let Q_ω be an r.c.p.d. for $E(\cdot|F_{T_1})$, let $X_t^{(1)} = C_1 X_{t+T_1}$, and let $\sigma_t^{(1)}$ be the diffusion coefficient of $X_t^{(1)}$. Using Ito's formula, we check that $\sigma_t^{(1)} = C_1 \sigma_{t+T_1} C_1$. Also $\sigma_0^{(1)} = I$.

Let $f_b(y) = \mu * \phi_b(C_1 y)$. Applying T2 to the process $\hat{X}_t^{(1)}$ corresponding to $\Gamma(\sigma_t^{(1)})$, $\int_0^t f_b(X_s^{(1)})ds$ converges uniformly for $t \le T_2 = \inf\{t: |\sigma_t^{(1)} - I| > \varepsilon\} \wedge (\tau - T_1)$, a.s. (dQ_ω).

Note again that

$$Q_\omega(|A_{T_1+T_2}(\mu * \phi_b) - A_{T_1}(\mu * \phi_b)|^p) = Q_\omega\left(\left|\int_0^{T_2} f_b(\hat{X}_s^{(1)})ds\right|^p\right)$$

$$\le c_{7.1}(p),$$

and similarly for conditional expectations.

Thus, $A_t(\mu * \phi_b)$ converges uniformly for $t \le T_1 + T_2$ and

$$E\, A^p_{T_1+T_2}(\mu * \phi_b) \le c_{7.2}(p).$$

Continue by induction: $X_t^{(2)} = C_2 X_{t+T_1+T_2}$, where C_2 is a square root of $\sigma^{-1}_{T_1+T_2}, \ldots$. Letting $V_n = T_1 + \cdots + T_n$, we get

(i) $A_t(\mu * \phi_b)$ converges uniformly for $t \le V_n$, and

(ii) $E\, A^p_{V_n}(\mu * \phi_b)) \le c_{7.3}(p,n)$ (and similarly for conditional expectations).

Using the first of these two facts, we have T1 for $t \le V_n$. If we let the stopping time T in the proofs of T3, T4 be $V_n \wedge \inf\{t: |X_t| \ge M\}$ and also use (ii), we have the results of T3, T4 for $t \le V_n$. We can get T2 for $t \le V_n$ in an exactly similar fashion. To finish the proof, it remains to show $V_n(\omega) = \tau$ for some n on, a.s.

Recall that $T_{n+1} = \inf\{t: |\sigma_t^{(n)} - I| > \varepsilon\} \wedge (\tau - V_n)$, and $\sigma_t^{(n)} = C_n\sigma_{t+V_n}C_n$, C_n a square root of $\sigma_{V_n}^{-1}$. We then get $|C_n\sigma_{V_{n+1}}C_n - I| = \varepsilon$ on the set $V_{n+1} < \tau$.

Let $V = \sup_n V_n$. Recall $\mathrm{trace}(K_1K_2) = \mathrm{trace}(K_2K_1)$. By the continuity of σ, $\sigma_{V_n} \to \sigma_V$, and so

$$\mathrm{trace}(C_n\sigma_{V_{n+1}}C_n - I)^2 = \mathrm{trace}(C_n^4\sigma_{V_{n+1}}^2) - 2\,\mathrm{trace}(C_n^2\sigma_{V_{n+1}}) + d$$

$$\to \mathrm{trace}((\sigma_V^{-1})^2\sigma_V^2) - 2\,\mathrm{trace}(\sigma_V^{-1}\sigma_V) + d = 0.$$

We must therefore have $V_{n+1}(\omega) = \tau$ for some n on, as required. □

Next we allow there to be drift.

REMOVAL OF R2. Define a probability Q on $C[0,\infty)$ by

$$(dQ/dP)|_{F_t} = \exp(M_t - \tfrac{1}{2}\langle M,M\rangle_t),$$

where $M_t = -\int_0^t (\sigma_t^{-1}\eta_t)^* dW_t$.

If $\hat{W}_t = \int_0^t \sigma_t^{-1} dX_t$, then under Q, $\hat{W}_t$ will be a Brownian motion. Under Q, (X,σ) satisfy A2 and R2, hence $\sup_{t\le\tau}|A_t(\mu*\phi_b) - A_t(\mu)| \to 0$, a.s. (dQ). Since P is absolutely continuous with respect to Q, this limit is 0, a.s. (dP). This proves T1, and T2 follows analogously.

T3 and T4 would follow as in section 6, provided we had bounds on $E_P(|A_T(\mu*\phi_b) - A_t(\mu*\phi_b)|^p|F_t)$, $t \le T$, T a stopping time as in the proofs of T3, T4. Such bounds are obtained by applying the following lemma, letting $R = P_\omega$, an r.c.p.d. for $E_P(\cdot|F_t)$ and $L = A_T(\mu*\phi_b) - A_{t\wedge T}(\mu*\phi_b)$.

(7.1) *LEMMA. Let* R *be a probability,* W_t *a* d-*dimensional Brownian motion,* $a_t(\cdot)$ *an* $\mathbb{R}^{d\times d}$ *valued functional on* $C[0,t]$, b_t *a predictable process. Suppose* $|a_t^{-1}| \le c_{7.4}$, $|b_t| \le c_{7.5}$.

Suppose Y_t *is a solution to* $Y_0 = y_0$, $dY_t = a_t(Y)dW_t + b_t dt$.

Suppose Z_t *is the unique (in law) solution to* $Z_0 = y_0$, $dZ_t = a_t(Z)dW_t$. *Let* L *be a functional on* $C[0,\tau]$. *Then* $E_R|L(Y)|^p \le c_{7.6}(E_R|L(Z)|^{2p})^{\frac{1}{2}}$, $c_{7.6}$ *depending only on* p, $c_{7.4}$, $c_{7.5}$, *and* τ.

PROOF. Define a probability Q on $C[0,\infty)$ by

$$(dQ/dR)|_{F_t} = N_t = \exp(M_t - \tfrac{1}{2}\langle M,M\rangle_t),$$

where $M_t = -\int_0^t (a_t^{-1}(Y)b_t)^* dW_t$. Under Q, $\hat{W}_t = \int_0^t a_t^{-1}(Y)dY_t$ is a Brownian motion, or $dY_t = \int_0^t a_t(Y)d\hat{W}_t$. By the uniqueness in law of Z,

$$E_Q|L(Y)|^{2p} = E_R|L(Z)|^{2p}.$$

Then

$$\begin{aligned} E_R|L(Y)|^p &= E_Q(|L(Y)|^p(dR/dQ)) \\ &\le E_Q(|L(Y)|^{2p})^{\frac{1}{2}}(E_Q N_\tau^{-2})^{\frac{1}{2}}. \end{aligned}$$

The result follows from standard estimates on $E_Q(N_\tau^{-2}) = E_R(N_\tau^{-1})$. □

Finally, we remove the restriction $d \ge 3$.

REMOVAL OF R3. Given a measure μ on $\mathbb{R}^d$, $d \le 2$, define $\mu\hat{}(C \times D) = \mu(C)\lambda|_{B(M,0)}(D)$, where $C \subseteq \mathbb{R}^d$, $D \subseteq \mathbb{R}^2$, and λ is Lebesgue measure on $\mathbb{R}^2$. Let $\hat{X}_t = (X_t, \hat{W}_t)$, where $\hat{W}_t$ is a 2-dimensional Brownian motion independent of the Brownian motion in terms of which X is defined.

If $t \le T = \inf\{t: |X_t| \ge M\}$, $X_t \in C$ if and only if $\hat{X}_t \in C \times B(M,0)$. It is not hard to see that $A_t((\mu * \phi_b)\hat{})$ has a uniform limit, $t \le T$, $b \to 0$ using T2. Here ϕ_b is the density of a d-dimensional normal random variable. Call this limit $A_t(\mu)$. T2 may be proved in the same fashion.

$E\, A_t^p((\mu * \phi_b)^\wedge) \le c_{7.7}(p)$ because $(\mu * \phi_b)^\wedge$ satisfies M1. Using this, together with the analogous statement for conditional expectations, we can prove T3 and T4 as in section 6. □

8.0 An Alternate Approach

If one merely wants to construct an occupation time whose support is a given smooth surface, there is a simpler way to proceed. Of course, the results obtained are much weaker also.

Let C be all or part of a C_2 surface: suppose there exists $f\colon \mathbb{R}^d \to \mathbb{R}^d$ such that $f \in C^2$, the Jacobian of f is bounded away from 0, and $f(C) = \bar{B} \times \{0\}$, B an open (possible unbounded) subset of $\mathbb{R}^{d-1}$, $\bar{B}$ = closure of B.

Suppose X_t is an $\mathbb{R}^d$-valued process, $dX_t = \sigma_t dW_t + \eta_t dt$, where σ_t, η_t are bounded and predictable, σ_t is uniformly strictly elliptic, and σ_t is continuous in t.

(8.1) *THEOREM. There exists a continuous increasing process* $\hat{A}_t$ *such that*

(i) *the support of* $\hat{A}_t = C$*, i.e.,* C *is the smallest closed set for which* $\int 1_{C^c}(X_s)d\hat{A}_s = 0$;

(ii) *there exist nonnegative functions* g_n *such that* $\hat{A}_t$ *is the uniform limit of* $\int_0^t g_n(X_s)ds$, $t \le \tau$, a.s.

It will be apparent from the proof that the conditions on X and C can be slightly weakened by a localization argument. Also, if one has a family of curves C_a that vary smoothly in a: $f(C_a) = \bar{B} \times \{a\}$, then $\hat{A}_t^a$

may be chosen to vary smoothly in a.

First we consider the one-dimensional case.

(8.2) *LEMMA. Suppose* $M_t = \int_0^t \sigma_s dW_s$, $d = 1$, σ_s *continuous in* s, $\sigma_s \geq c_{8.1}$ *for all* s, a.s. *Then there exist versions of* $A_t^\varepsilon = (2\varepsilon)^{-1} \int_0^t 1_{[-\varepsilon,\varepsilon]}(M_s)ds$ *that converge uniformly*, $t \leq \tau$, a.s. *to an increasing process whose support is* $\{0\}$.

PROOF. First of all, if g_ε is given by $g_\varepsilon(0) = 0$, $g_\varepsilon'(x) = (\min(1,|x|/\varepsilon))\mathrm{sgn}(x)$, Ito's formula gives

$$L_t^\varepsilon = (2\varepsilon)^{-1} \int_0^t 1_{[-\varepsilon,\varepsilon]}(M_s)d\langle M,M\rangle_s = g_\varepsilon(M_t) - g_\varepsilon(M_0) - \int_0^t g_\varepsilon'(M_s)dM_s.$$

By a familiar argument, L_t^ε converges uniformly in $t \leq \tau$, a.s. to L_t, the local time of M at 0. (One estimates $E|\int_0^t (g_\varepsilon' - g_\delta')(M_s)dM_s|^p$ by Burkholder's inequality, and then uses Kolmogorov's theorem.)

If H_s is continuous and bounded, $\int_0^t H_s dL_s^\varepsilon$ converges uniformly, $t \leq \tau$, to $\int_0^t H_s dL_s$, a.s. This may be proved by fixing ω, and then approximating $H_s(\omega)$ by a step function $H_s^n = \sum_{i=1}^n k_i 1_{[s_i,s_{i+1})}(s)$ so that $\sup_{s \leq \tau}|H_s^n - H_s| < \delta$.

Now let $H_s = \sigma_s^{-1}$, let $A_t = \int_0^t H_s dL_s$, and observe that $A_t^\varepsilon = \int_0^t H_s dL_s^\varepsilon \to A_t$. □

It is known that versions of L_t^x, the local time at x, can be chosen that are jointly continuous in x and t. It follows that $A_t^x = \int_0^t H_s dL_s^x$ can be taken to be jointly continuous in x and t.

PROOF of (8.1). By localization, let us suppose $\|D_{ij}f\|$, $\|D_i f\|$, $\|f\|$ are all bounded. Let $Y_t = f(X_t)$. Using Ito's lemma, Y_t is a d-dimensional semimartingale satisfying

$$dY_t^k = \sum_i D_i f^k(X_t)dX_t^i + \frac{1}{2}\sum_{i,j} D_{ij}f^k(X_t)d\langle X^i,X^j\rangle_t.$$

Here f^k is the kth coordinate of f.

Let $J_{ik}(t) = D_i f^k(X_t)$. Note that the drift coefficients of Y, $\frac{1}{2}\sum D_{ij}f^k(X_t)\sigma_{ij}(t)$, are bounded.

The diffusion coefficients of Y_t are given by

$$d\langle Y^k,Y^\ell\rangle_t = \langle \sum_i J_{ik}(t)dX_t^i, \sum_j J_{j\ell}dX_t^j\rangle = (J_t^*\sigma_t J_t)_{k\ell}dt.$$

Since σ is strictly elliptic and symmetric, $J_t^*\sigma_t J_t$ is nonnegative definite and will be strictly elliptic provided $\det(J^*\sigma J) > c_{8.2} > 0$. This follows from the hypothesis that the Jacobian of f is bounded away from 0.

Thus the problem is reduced to the case where $C = \bar{B}\times\{0\}$, B an open subset of $\mathbb{R}^{d-1}$ and Y_t satisfies the same conditions at X_t. Arguing as in section 7, we may assume that Y_t has 0 drift.

By (8.2),

$$A_t^\varepsilon = \int_0^t (2\varepsilon)^{-1}\, 1_{\mathbb{R}^{d-1}\times[-\varepsilon,\varepsilon]}(Y_s)ds$$

$$= \int_0^t (2\varepsilon)^{-1}\, 1_{[-\varepsilon,\varepsilon]}(M_s)ds \to A_t,$$

where M_s is the dth coordinate of Y_s.

Let $h\colon \mathbb{R}^d \to \mathbb{R}$ be a continuous bounded function whose support is exactly $\bar{B}\times\mathbb{R}$ and such that $h(y_1,\ldots,y_d)$ does not depend on y_d. Then

$$\hat{A}_t^\varepsilon = \int_0^t (2\varepsilon)^{-1}\, 1_{\mathbb{R}^{d-1}\times[-\varepsilon,\varepsilon]}(Y_s)h(Y_s)ds$$

$$= \int_0^t h(Y_s)dA_s^\varepsilon \to \int_0^t h(Y_s)dA_s = \hat{A}_t,$$

as in this proof of (8.2).

To finish the proof, we must show that $f(C)$ is the support of $\hat{A}$. By the definition of $\hat{A}$, support $(\hat{A})$ = support (h) $\cap$ support (A). Since $d\langle M,M\rangle_t/dt$ is bounded above and below away from 0, support (A) = support (L) = $\mathbb{R}^{d-1} \times \{0\}$. And $(\bar{B} \times \mathbb{R}) \cap (\mathbb{R}^{d-1} \times \{0\}) = \bar{B} \times \{0\} = f(C)$. □

(8.3) *EXAMPLE.* Let $C_u = \{(t,f(t) + u): 0 \le t \le 1\}$, where $f(t)$ is a continuous, nowhere differentiable function (e.g., a typical Brownian path). Obviously (8.1) does not apply to C_u.

Let $\mu_u(B)$ = (one-dimensional) Lebesgue measure of $\{t: (t,f(t) + u) \in B\}$. Clearly $\mu_u(B(r,x)) \le 2r$.

By T1, $A(\mu_u)$ exists. It is conceivable that $A_1(\mu_u)$ is 0 for some u. (This would depend on the process X_s; it would be impossible for it to occur for a two-dimensional Brownian motion, for example.) However, for almost all u, $A_1(\mu_u)$ must be nonzero.

References

1. R. F. BASS. Joint continuity and representations of additive functionals of d-dimensional Brownian motion, to appear.

2. J. M. BISMUT. Martingales, the Malliavin calculus and hypoellipticity under general Hörmander's conditions, *Z. Wahrscheinlichkeitstheorie verw. Gebiete 56* (1981), 469-505.

3. A. FRIEDMAN. *Stochastic Differential Equations and Applications*, vol. 1. Academic Press, New York, 1975.

4. U. G. HAUSSMANN. On the integral representation of functionals of Itô processes. *Stochastics 3* (1979), 17-27.

5. N. IKEDA and S. WATANABE. *Stochastic Differential Equations and Diffusion Processes*. North-Holland/Kodansha, New York, 1981.

6. N. V. KRYLOV. *Controlled diffusion processes.* Springer-Verlag, New York, 1980.

7. B. MAISONNEUVE. Ensembles régenératifs, temps locaux et subordinateurs. Séminaire de Probabilités V (*Univ. Strasbourg*), pp. 147-169. Lecture Notes in Math. *191*, Springer-Verlag, Berlin, 1971.

8. P. A. MEYER. Un cours sur les intégrales stochastiques. Séminaire de Probabilités X. (*Univ. Strasbourg*), pp. 245-400. Lecture Notes in Math. *511*, Springer-Verlag, Berlin, 1976.

9. W. RUDIN. *Real and Complex Analysis.* McGraw-Hill, New York, 1966.

10. D. W. STROOCK and S. R. S. VARADHAN. *Multidimensional Diffusion Processes.* Springer-Verlag, New York, 1979.

R. F. BASS
Department of Mathematics
University of Washington
Seattle, Washington 98195

Seminar on Stochastic Processes, 1982
Birkhäuser, Boston, 1983

A SIMPLE VERSION OF THE MALLIAVIN CALCULUS IN DIMENSION N

by

KLAUS BICHTELER and DAVID FONKEN

1. Introduction and Dependence of Solutions on Parameters

The solution X of a Lipschitz stochastic differential equation depends smoothly on parameters. In conjunction with Girsanov's theorem this produces a simple and straightforward proof of the existence and smoothness of a density for the transition kernel of the Markov process X.

Consider the stochastic differential equation

$$dX_t^x = a(X_t^x)dW_t + b(X_t^x)dt, \quad X_0^x = x, \tag{1.1}$$

where W is an M-dimensional standard Wiener process, $a(x)$ is an $N \times M$-matrix depending on $x \in R^N$, and b is a vector field on R^N. We shall assume throughout that a and b are Lipschitz:

$$|a(x) - a(y)| \le K|x - y| \quad \text{and} \quad |b(x) - b(y)| \le K|x - y|.$$

Norms denoted by $|\cdot|$ are understood in the sense of ℓ_2. There is a unique solution to (1.1), which is a Markov process with transition function P_t given by

$$P_t f(x) = Ef(X_t^x), \quad f \in C_b(R^N).$$

By studying the partial differential equation satisfied by $u(t,x) = P_t f(x)$, Hörmander [H1] showed in 1967 that, under suitable smoothness assumptions on a and b, the transition function P_t has a smooth density p_t:

$$(1.2) \qquad P_t f(x) = \int f(y) p_t(x,y) dy.$$

Now (1.2) is an easy consequence of the estimates

$$(1.3) \qquad |E[\partial^k/(\partial x)^k f(X_t^x)]| \le C_k \| f \|_\infty, \quad f \in C^\infty(R^N),$$

which imply that the Fourier transform of P_t decreases faster than any polynomial at infinity.

For long years it was an embarrassment to the probabilistic community that the purely probabilistic conclusion (1.2) about the probabilistic situation (1.1) could only be proved by Hörmander's analytic methods. In 1974, Malliavin rectified this situation with his variational calculus on Wiener space. The latter was streamlined by Stroock [S1] and Bismut [B3]; nevertheless, the Malliavin Calculus remains an apparatus of formidable technical aspect. Our purpose is to show that enlarging the scope slightly reduces the arguments to rather straightforward computations and estimates involving not more than Girsanov's theorem. The procedure is rather close to Bismut's.

Once one question about smoothness of P_t is raised, others come to mind immediately; for instance, does $P_t f$ depend smoothly on x? Or does the process X depend smoothly on parameters that influence a and b? The key to our approach is the answer to the last question, and this answer is precisely what one expects by analogy with stability results from the theory of Banach space valued ODE's as, for example, in Dieudonné [D1].

The stability results concerning the stochastic equation (1.1) with proofs in all detail and generality have appeared in [B2]. Here, we shall give a description of both results and methods of proof, in order to convey the idea that it is all rather straightforward.

Usually (1.1) is solved by finding a fixed point for the map T from (adapted, continuous) processes to processes defined by

$$T(X)_t = x + \int_0^t a(X_s)dW_s + \int_0^t b(X_s)ds.$$

In fact, it is easily seen that for any $p \geq 2$ there is an $\alpha = \alpha(p)$ so that T is contractive on the Banach space $B_{p,\alpha}$ of processes X whose norm

$$\| X\|_{p,\alpha} = \sup_{t \geq 0} e^{-\alpha t} \| |X|_t^* \|_{L^p}$$

where

$$X_t^* = \sup_{s \leq t} |X_s|,$$

is finite.

Suppose now that x, a, and b depend on some parameter $u \in R^N$. Then so will $T = T(u)$ and the fixed point $X = X(u)$. Recall what it means that X depends differentiably on u at u: there is a linear map $DX(u)\colon R^N \to B_{p,\alpha}$ and a remainder R with

$$\| R(v,u)\|_{p,\alpha} = o(|v-u|)$$

so that

$$X(v) - X(u) = DX(u)(v-u) + R(v,u).$$

THEOREM. *Suppose* a(x,u), b(x,u) *and* x(u) *have bounded continuous derivatives up to order two. Then* x(u) *is differentiable and the derivative solves the equation obtained by formal derivation of*

$$X_t(u) = x(u) + \int_0^t a(X(u),u)dW + \int_0^t b(X(u),u)ds \tag{1.4}$$

so that

$$\begin{aligned} DX_t(u) = {} & \left\{Dx(u) + \int_0^t \partial a/\partial u(X(u),u)dW + \int_0^t \partial b/\partial u(X(u),u)ds\right\} \\ & + \int_0^t \partial a/\partial x(X(u),u)DX(u)dW + \int_0^t \partial b/\partial x(X(u),u)DX(u)ds. \end{aligned} \tag{1.5}$$

The way to prove this is to define $DX(u)$ as the solution to (1.5) and to check that $R(v,u) = X(v) - X(u) - DX(u)(v-u)$ satisfies a linear differential equation with "initial condition" $= o(|v-u|)$. This forces $\|R(v,u)\|_{p,\alpha} = o(|v-u|)$. Now, if a and b are smooth enough, one can differentiate again, etc. The situation is complicated somewhat by the fact that terms like $\int \partial^2 a/\partial x^2(X(u),u)DX(u) \otimes DX(u)dW$ appear, which will lie in $B_{p/2,\alpha}$ but not in $B_{p,\alpha}$. That is, degrees of integrability are lost as higher derivatives are taken, but this little problem can be overcome easily enough by simply starting with sufficiently large p. There is another complication stemming from the fact that $f \circ X: B_{p,\alpha} \to B_{p,\alpha}$ is not differentiable in the ordinary sense even if f has all bounded derivatives, but that can be overcome as well. Also, the theorem cannot be applied as stated to (1.5) to obtain the second derivative, because the "initial condition $\{\cdots\}$" is not constant and

$$DX \to \partial a/\partial x(X(u),u)DX$$

is not of the simple form $X \to A(X,u)$. But these obstacles are easily overcome by suitable generalizations, and the upshot is still what one expects:

THEOREM. *Suppose* a *and* b *have* $k+1$ *bounded derivatives in* (x,u). *Then for any* $p \geq 2$ *there is an* $\alpha = \alpha(p)$ *so that the map* $X: u \to X(u)$ *from* R^N *to* $B_{p,\alpha}$ *defined by* (1.4) *is* k-*times differentiable, the derivatives being the solutions of the differential equations obtained by successive formal differentiation of* (1.4). *In particular,* $|D^{(i)}X(u)|_t^* \in L^p$ *for all* p *and* $0 \leq |i| \leq k$.

2. The Existence and Smoothness of Densities

We wish to apply the results of the previous section to show that the solution X_t of equation (2.1) listed below has under certain conditions a smooth density. Our starting point is the following fundamental lemma; see [S1] for its proof.

LEMMA. *Let* μ *be a finite Radon measure on* R^N *and assume that there is a* $C < \infty$ *and an* $n > N$ *such that* $|\int f^{(\alpha)}(y)\mu(dy)| \leq c\cdot\sup_y|f(y)|$, $f \in C_b^\infty(R^N)$, *for all* $|\alpha| \leq n (f^{(\alpha)} \equiv D^{(\alpha)}f(x))$. *Then* μ *is absolutely continuous with respect to Lebesgue measure; and if* $p = d\mu/dx$, *then* $p \in C_b^k(R^N)$ *where* $k = n - N - 1$ *and* $\|p\|_{C_b^k(R^N)} \leq A(n,N)\cdot C$ *where* $A(n,N)$ *depends on* n *and* N *alone.*

Let $W_t = (W_t^i)^{i=1,\ldots,M}$ be a standard Brownian motion on a stochastic basis $(\Omega, (F_t)_{0\leq t\leq\tau}, P)$. $S^p(P)$ will denote the space of continuous processes Y on $[0,\tau]$ with norm the $L^p(P)$-norm of the maximal function Y_t^*. Suppose that $a = (a_j^i)_{j=1,\ldots,M}^{i=1,\ldots,N}$ and $b = (b_j)_{j=1,\ldots,N}$ are such that $a_j^i, b_j: R^N \to R$ have bounded derivatives of all orders. We will investigate how the solution of the equation

$$dX_t = a(X_t)dW_t + b(X_t)dt; \quad X_0 = x \tag{2.1}$$

can be perturbed via Girsanov's theorem.

Girsanov's theorem tells us that under suitable conditions on the process $H_t = (H_j^i(t))_{j=1,\dots,N}^{i=1,\dots,M}$ the process $W_t(u) = (W_t^i(u))^{i=1,\dots,M}$, where $u \in R^N$, defined by

$$W_t(u) = W_t - \int_0^t H_s u \, ds \tag{2.2}$$

is a standard $((\mathcal{F}_t), P(u))$ Brownian motion. The measure $P(u)$ is defined by $dP(u) = G(H,u)dP$ where $G_t(H,u) = \exp(\int_0^t H_s u \, dW_s - \frac{1}{2}\int_0^t \|H_s u\|^2 ds)$ is the solution to $dG_t = H_t u G_t dW_t$; $G_0 = 1$.

Consider the perturbed equation

$$dX_t(u) = a(X_t(u))dW_t + [b(X_t(u)) - a(X_t(u))H_t u]dt; \quad X_0(u) = x. \tag{2.3}$$

Note that when u is 0 this is simply (2.1). Noting (2.2) we obtain

$$dX_t(u) = a(X_t(u))dW_t(u) + b(X_t(u))dt; \quad X_0(u) = x. \tag{2.4}$$

Compare equations (2.1) and (2.4). The solutions are not the same but since the parameter u does not appear explicitly in (2.4), the $P(u)$-distribution of $X_t(u)$ must be the P-distribution of X_t for every choice of u in R^N.

We now apply the results of section 1: $X_t(u)$ depends differentiably on u and at $u = 0$ the matrix of derivatives

$$D_t := \left.\frac{\partial X_t}{\partial u}\right|_{u=0}$$

must satisfy the equation obtained from (2.3) by formal differentiation:

$$(2.5)\qquad dD_t = \frac{\partial a_i}{\partial x}(X_t)D_t dW_t^i + \frac{\partial b}{\partial x}(X_t)D_t dt + a(X_t)H_t(W); \quad D_0 = 0.$$

A few comments are in order here: First, a_i denotes the ith column of the matrix a, so that $\frac{\partial a_i}{\partial x}$ is a three index tensor. Second, the Einstein summation convention is being employed. Finally, we have written H_t as $H_t(W)$ to indicate that it is assumed from this point on that the paths of H up to time t depend measurably on those of W up to time t, for every t (as might be the case if H_t were the solution of a stochastic differential equation driven by W_t [B1]).

Let us modify (2.5) to define a matrix of processes $D_t(u)$ as the solution to

$$(2.6)\qquad dD_t(u) = \frac{\partial a_i}{\partial x}(X_t(u))D_t(u)dW_t^i(u) + \frac{\partial b}{\partial x}(X_t(u))D_t(u)dt$$
$$+ a(X_t(u))H_t(W(u)); \quad D_0(u) = 0.$$

Just as in (2.4) the P(u)-distribution of $D_t(u)$ does not depend on u.

For the sake of notational convenience let us define an operator L by $L(X_t(u)) := D_t(u)$. L can be applied to the solution of any stochastic differential equation driven by $W_t(u)$ to which the theorem of section 1 applies. In every case

$$LX_t(0) = \left.\frac{\partial X_t}{\partial u}\right|_{u=0}$$

and the P(u)-distribution of $LX_t(u)$ will be independent of u.

If we define $D_t^k(u) := LD_t^{k-1}(u)$ for $k > 1$, then

$$(2.7)\qquad dD_t^k(u) = \frac{\partial a_i}{\partial x}(X_t(u))D_t^k(u)dW_t^i(u) + \frac{\partial b}{\partial x}(X_t(u))D_t^k(u)dt$$
$$+ p_t(u)dW_t(u) + q_t(u)dt; \quad D^k(u) = 0,$$

where the components of $p_t(u)$ and $q_t(u)$ are polynomials in the components of $D_t(u)$, $(\frac{\partial}{\partial u})^{j-1}\ a(X_t(u))H_t(W(u))$, and $(\frac{\partial}{\partial x})^j\ a(x)$ and $(\frac{\partial}{\partial x})^j\ b(x)$ evaluated at $X_t(u)$ for $j < k$.

We will also apply the operator L to $G_t(u)$ which appeared when we applied Girsanov's theorem to equation (2.1) and satisfies

$$\text{(2.8)} \qquad dG_t(u) = G_t(u)\sigma(H_t(W)u)H_t(W)udW_t; \quad G_0(u) = 1.$$

Now suppose g is a polynomial in the components of $D_t(u)$, $D_t^2(u)$, $\ldots, D_t^k(u)$ and $LG_\tau(u), \ldots, L^kG_\tau(u)$, ψ is a function of the components of $\bar{D}_t(u) = D_t^{-1}(u)$ and f is a smooth function, then

$$\text{(2.9)} \qquad E[f(X_t(u))\cdot\psi\cdot g\cdot G_\tau(u)] = E^u[f(X_t(u))\cdot\psi\cdot g]$$

does not depend on u. The partial derivative with respect to u^i at $u = 0$ must be zero. Applying Leibnitz' rule and using $X_t = X_t(0)$, $D_t^k = D_t^k(0)$, $g = g(0)$ and $G_\tau(0) = 1$ we obtain:

$$\text{(2.10)} \qquad E\Big[\frac{\partial f}{\partial x_j}(X_t)D_i^j(t)\cdot\psi\cdot g\Big] = -E\Big[f(X_t)\cdot\frac{\partial}{\partial u^i}\{\psi\cdot g\cdot G_\tau(u)\}\big|_{u=0}\Big].$$

Now choose $\psi = \bar{D}_k^i$ and sum over the index i. The result is:

$$\text{(2.11)} \qquad E\Big[\frac{\partial f}{\partial x_k}(X_t)\cdot g\Big] = -E[f(X_t)H_k[g]]$$

where $H_k[g]$ is of the same form as g, except that $|\alpha| \leq n+1$. More generally,

$$\text{(2.12)} \qquad E[f^{(\alpha)}(X_t)] = (-1)^{|\alpha|}E[f(X_t)H^\alpha[1]]$$

where $H^{\alpha} = H_1^{\alpha_1} \circ \cdots \circ H_N^{\alpha_N}$. Taking $C^{\alpha} = \|H^{\alpha}[1]\|_{L^p}$ we obtain

$$(2.13) \qquad \left|\int f^{(\alpha)}(y)p_t(x,dy)\right| \le C^{\alpha}\cdot\sup_y|f(y)|.$$

It remains to show that the constants C^k are finite. Now $H^{\alpha}[1]$ is a polynomial in the components of $\bar{D}_t$, $D_t,\ldots,D_t^{k-1}$ and $LG_{\tau}(0),\ldots,L^{k-1}G_{\tau}(0)$ all of which will be in S^p for each $p < \infty$ by the results of section 1 (once H_t is chosen properly). Hölder's inequality will then yield $C^{\alpha}<\infty$.

There is actually a small problem with (2.13). To apply Girsanov's theorem we need a condition on H_t like $\int_0^t \|H_s u\|^2 ds < C < \infty$ and this will prove to be too restrictive. Define stopping times $T_n = \inf\{t > 0: \int_0^t \|H_s u\|^2 ds \ge n\}$ and let

$$H_t^n = \begin{cases} H_t & t \le T \\ 0 & t > T \end{cases}.$$

Replace H_t by H_t^n in the above argument. Equation (2.9) will now read: $E[f(X_t^n(u))g(n)\psi(n)G_{\tau}^n(u)]$ is independent of u. Let $0 \le \phi \le 1$ be a function on $R^N \otimes R^N$ such that

$$\phi(A) = \begin{cases} 1 & \text{if } \|A^{-1}\| \le \ell \\ 0 & \text{if } \|A^{-1}\| \ge 2\ell \text{ or } A \text{ is noninvertible} \end{cases}$$

and such that ϕ has bounded derivatives of all orders. Choose $\psi(n) = \psi\cdot\phi(D_t^n)$ where ψ, as before, is a function of the components of $\bar{D}_t^n(u)$. If we let $n \to \infty$, equation (2.9) tells us that $E[f(X_t(u))\cdot g\cdot\psi\cdot\phi(D_t)G_{\tau}^{\infty}(u)]$ is independent of u. There are now two cases to consider:

If $\bar{D}_t \in S^p$ for all $p < \infty$ then let $\ell \to \infty$ and proceed exactly as before to obtain (2.13).

If not, we cannot replace ϕ by 1, and obtain in place of (2.11):

$$E\left[\frac{\partial f}{\partial x_k}(X_t)\cdot\phi\cdot g\right] = -E[f(X_t)H_k[\phi\cdot g]] \tag{2.14}$$

and consequently in place of (2.13):

$$\left|E\left[\frac{\partial f}{\partial x_k}(X_t)\,\phi(D_t)\right]\right| \le C_k\cdot\sup_y|f(y)|. \tag{2.15}$$

We turn now to the problem of making the proper choice for the perturbing factor H_t. Let Y_t denote the solution to

$$dY_t = \frac{\partial a_i}{\partial x}\cdot Y_t\cdot dW_t^i + \frac{\partial b}{\partial x}\cdot Y_t\cdot dt; \quad Y_0 = I_N. \tag{2.16}$$

Both Y and its inverse $\bar{Y}$ (which satisfies a similar linear equation) belong to S^p for all $p < \infty$. Applying the method of variation of parameters to equation (2.7) yields:

$$D_t = Y_t\cdot\int_0^t \bar{Y}_s a_s H_s ds. \tag{2.17}$$

Suppose we choose $H_t = a^T(X_s)\bar{Y}_s^T$ so that $D_t = Y_t\cdot\int_0^t \bar{Y}_s a_s a_s^T \bar{Y}_s^T\, ds$. Then we have the following (Hörmander's conditions):

PROPOSITION. *If the span of* $a_1(x),\dots,a_M(x)$ (*where* a_i *denotes the* i^{th} *column vector of* a); $[a_i, a_j],[b,a_i]_{1\le i,j\le M}$; $[a_j, [a_i, a_k]]$, $[b,[a_j, a_k]]$, $[a_i, [b, a_k]]_{1\le i,j,k\le M}$;... *is all of* $T_x(R^N)$ *then* D_t *is a.s. invertible.*

PROOF: The following argument is due to Bismut [B3]. Let U_s be the span of $\{\bar{Y}_s a_i\colon i=1,\dots,M\}$, V_t be the span of $\bigcup_{s\le t} U_s$ and $V_t^+ = \bigcap_{s\le t} V_s$.

By the Blumenthal zero-one law, V_0^+ a.s. does not depend on $\omega \in \Omega$.

Suppose $V_0^+ \neq T_x(R^N)$. Then if $S = \inf\{t > 0: V_t \neq V_0^+\}$, a.s. $S > 0$. Let $f \in T_x^*(R^N)$ be orthogonal to V_0^+. Then

$$\langle f, \bar{Y}_t a_i \rangle = 0 \quad \text{for } t \leq S,\ 1 \leq i \leq M. \tag{2.18}$$

Applying Ito's formula to $\bar{Y}_t a_i$ we obtain

$$0 = \langle f, \bar{Y}_t a_i \rangle = \langle f, a_i(x) \rangle + \int_0^t \langle f, \bar{Y}[a_j, a_i] \rangle dW^j$$
$$+ \int_0^t \langle f, \bar{Y}_s([b, a_i] + \tfrac{1}{2}[a_j,[a_j,a_i]]) \rangle ds.$$

Consequently

$$\langle f, \bar{Y}_t[a_j, a_i] \rangle = 0 \qquad 1 \leq i,j \leq M, \quad t \leq S. \tag{2.19}$$

Repeating the above procedure with (2.19) gives

$$\langle f, \bar{Y}_t[a_j,[a_j,a_i]] \rangle \qquad t \leq S$$

so that

$$\langle f, \bar{Y}_t[b, a_i] \rangle = 0 \qquad t \leq S.$$

Iterating the above procedure we find that for any of the brackets listed in the statement of this proposition that $\langle f, \bar{Y}_t[\cdots] \rangle = 0$, $t \leq S$. Taking $t = 0$ and noting that $\bar{Y}_0 = I_N$ we find that f is orthogonal to $T_x(R^N)$, a contradiction. □

This proposition tells us general conditions under which $\bar{D}_t$ is defined but not if $\bar{D}_t \in S^p$ for all $p < \infty$. Assuming the worst we get only inequality (2.15). This suffices to show that the distribution of X_t has a density but does not yield smoothness results.

Let us consider a special case where stronger hypotheses hold: Suppose that the matrix a is invertible at the starting point $x = X_0$. By continuity, we may choose a $c > 0$ such that $a(y)$ is invertible for every $y \in B_c(x)$. Define a stopping time $T = \inf\{t > 0: X_t \notin B_c(x)\}$ and choose in equation (2.2)

$$H_t = \begin{cases} \bar{a}_t Y_t & t \le T \\ 0 & t > T \end{cases} .$$

Then from (2.17) $D_t = Y_t \cdot T \wedge t$. To show that $\bar{D}_t \in S^p$ it suffices to show $\| \frac{1}{T} \|_{L^p} < \infty$. We apply Girsanov's theorem one final time:

Define $W_t' = W_t - \int_0^t h_s ds$ where

$$h_t = \begin{cases} -\bar{a}(X_t)b(X_t) & t \le T \\ 0 & t > T \end{cases}$$

and a measure P' by $dP' = \Gamma_T dP$ where

$$\Gamma_t = \exp\left(\int_0^t h_s dW_s - \frac{1}{2} \int_0^t \| h_s \|^2 ds\right).$$

By Girsanov's theorem (W_t', F_t, P') is a standard Brownian motion and equation (2.1) can be rewritten as

$$dX_t = a(X_t)dW_t' + b'(X_t)dt; \quad X_0 = x$$

where $b'(X_t) = b(X_t)\cdot[t > T] = 0$ when $t \le T$. Then X^T is a (P', F_t)-martingale. Since both Γ_T and $1/\Gamma_T$ are in L^p for all $p < \infty$ it suffices to show instead that $\|1/T\|_{L^p(P')} < \infty$ for all p.

We apply a recent result of Gundy's: For a martingale M with $M_0 = 0$, which has maximal function M^* and square function $[M,M]^{\frac{1}{2}}$ at $t = \infty$

$$\|(M^*)^3/[M,M]\|_{L^p} \leq c_p \|M^*\|_{L^p}. \tag{2.20}$$

By considering $X_t - x$ if necessary we may assume without loss of generality that $x = X_0 = 0$. Now $X_t = (X_t^i)^{i=1,\ldots,N}$ so let $M = (X^1)^T$. We may assume $|M|^* = c$ and $[M,M] = \int_0^{t \wedge T} \sum_j (a_j^1)^2 ds \leq N \cdot A^2 \cdot T$ where $A = \sup\{|a_j^i(y)| : 1 \leq i,j \leq N,\ y \in B_c(x)\}$. Now (2.20) gives

$$\|c^3/\{NA^2T\}\|_{L^p(P')} \leq c_p \cdot c$$

so $\frac{1}{T} \in L^p(P')$ hence $\frac{1}{T} \in L^p(P)$.

Inequality (2.13) holds in this case, and the existence of a C^∞ density for the distribution of X_t follows from the lemma stated at the beginning of this section.

References

[B1] K. BICHTELER. Stochastic integration and L^p-theory of semimartingales. *Ann. Prob.* *9* (1981), 49-89.

[B2] K. BICHTELER. Stochastic integrators with stationary independent increments. *Z. Wahrscheinlichkeitstheorie verw. Gebiete* *58* (1981), 529-548.

[B3] J.M. BISMUT. Martingales, the Malliavin calculus and hypoellipticity under general Hörmander's conditions. *Z. Wahrscheinlichkeitstheorie verw. Gebiete* *56* (1981), 469-505.

[D1] J. DIEUDONNÉ. *Foundations of Modern Analysis.* Academic Press, New York - London, 1960.

[H1] L. HÖRMANDER. Hypoelliptic second order differential equations. *Acta. Math. 119* (1967), 147-171.

[M1] P. MALLIAVIN. Stochastic calculus of variation and hypoelliptic operators. Proc. of the International Symposium on Stochastic Differential Equations (Kyoto 1976), Tokyo, 1978.

[S1] D. STROOCK. The Malliavin calculus and its application to second order parabolic differential equations: Part I. *Math. Systems Theory 14* (1981), 25-65.

KLAUS BICHTELER
Department of Mathematics
The University of Texas
Austin, Texas 78712

DAVID FONKEN
Department of Mathematics
The University of Texas
Austin, Texas 78712

Seminar on Stochastic Processes, 1982
Birkhäuser, Boston, 1983

AN INEQUALITY FOR BOUNDARY VALUE PROBLEMS*

by

K. L. CHUNG

Let D be a domain in R^d, $d \geq 1$, with $m(D) < \infty$ where m is the Lebesgue measure in R^d. The boundary ∂D of D is assumed to be smooth enough that the "area" measure σ is defined on it. For any positive (≥ 0) Borel measurable function $f \in L^1(\partial D, \sigma)$, namely integrable over ∂D with respect to σ, we define

$$(1) \qquad h(x) \equiv h(f;x) = E^x\{f(X(\tau_D))\}.$$

Here $X = \{X_t, t \geq 0\}$ is the Brownian motion process in R^d; E^x and P^x denote respectively the expectation and probability for the process starting at x, and

$$\tau_D = \inf\{t > 0 | X_t \notin D\}$$

is the first exit time from D. It is well known that if $h \not\equiv \infty$ in D, then h is harmonic in D. For this and other well-known results see [1]. In particular h is locally bounded, but not necessarily bounded, in D. Our first observation is that under general smoothness conditions

*This research is supported in part by NSF grant No. MCS-80-01540.

on ∂D, $h \in L^1(D,m)$, namely integrable over D with respect to m. Indeed, there exists a constant C depending only on D such that the inequality below holds for all positive $f \in L^1(\partial D,\sigma)$:

$$\int_D h(f;x)m(dx) \le C \int_{\partial D} f(y)\sigma(dy), \tag{2}$$

(to be abbreviated as $\int h\ dm \le C \int fd\sigma$). This inequality is a natural companion to the Harnack inequalities, but does not seem to have been recorded in the literature. Its proof under specific conditions will be postponed till the end of this paper. Our main purpose is to derive several interesting consequences from it, under whatever eventual assumptions which may be found to ensure its validity.

First we shall extend the inequality (2) to the case of solutions of Schrödinger equations with prescribed boundary values. This was studied in [3]; the basic notation used there will now be reviewed. Let q be a bounded Borel measurable function which is Hölder continuous in D, and let $Q = \sup_{x \in R^d} |q(x)|$. We shall reserve the symbol $\|\cdots\|$ for the sup-norm in D: $\|\phi\| = \sup_{x \in D} |\phi(x)|$. We put

$$e(t) = \exp\left(\int_0^t q(X(s))ds\right);$$

and for positive Borel measurable ϕ,

$$P_t\phi(x) = E^x\{\phi(X_t)\}, \qquad P_t^D\,\phi(x) = E^x\{t < \tau_D;\ \phi(X_t)\},$$

$$L_t\phi(x) = E^x\{t < \tau_D;\ e(t)\phi(X_t)\}.$$

For any positive Borel measurable f on ∂D, we put

(3) $$u(x) \equiv u(f;x) = E^x\{e(\tau_D)f(X(\tau_D))\}.$$

Thus when $q \equiv 0$, u reduces to the h in (1). When $f \equiv 1$ in (3), the corresponding $u(1;\cdot)$ is denoted by u_D in [3] and is called the "gauge" for (D,q). A basic result (Theorem 1.2 in [3]) states that if $u_D \not\equiv \infty$ in D, then u_D is bounded in $\bar{D}$. This lends a quick proof of the lemma below. From now on $C_1, C_2, \ldots$ will denote strictly positive constants which may depend on D and q.

LEMMA 1. *If* $u_D \not\equiv \infty$ *in* D, *then there exist* C_1 *and* C_2 *such that for all* $t \geq 0$

(4) $$\| L_t\, 1 \| \leq C_2\, e^{-C_1 t}.$$

PROOF. This can be proved using the (easy) result that all eigenvalues of the operator $\frac{\Delta}{2} + q$ are strictly negative, and standard spectral theory. (In [3] it is proved that the strict negativity of eigenvalues is equivalent to the hypothesis that $u_D \not\equiv \infty$ in D, under some smoothness condition on ∂D. If we assume that ∂D is merely regular in the sense of the Dirichlet boundary value problem, a detailed proof of (4) may be found in [5].) The following direct proof needs no condition on ∂D. The key lay hidden in the proof of Theorem 3.2 of [3], which can be readily strengthened, as shown me by Neil Falkner. To save the reader the trouble of digging it up, we repeat the argument here with a little simplification. For $t \geq 0$ we have

(5) $$E^x\{t < \tau_D \leq t+1;\, e(\tau_D)\} = E^x\{t < \tau_D;\, e(t)E^{X(t)}[e(\tau_D);\, 0 < \tau_D \leq 1]\}$$

$$\geq E^x\{t < \tau_D;\, e(t)\} \inf_{x \in D} P^x\{\tau_D \leq 1\} e^{-Q} = C_4 L_t\, 1(x).$$

The infimum above is strictly positive because $x \to P^x[\tau_D \le 1]$ is lower semi-continuous and strictly positive in D. The first member of (5) does not exceed $u_D(x)$, which belongs to $L^1(D,m)$ because u_D is bounded in D and $m(D) < \infty$. On the other hand, the said member converges to zero as $t \to \infty$. Hence we have by bounded convergence

(6) $$\lim_{t\to\infty} \int_D (L_t\, 1)dm = 0.$$

Now it is a plain but important property of L_t that for any $\phi \in L^1(D,m)$ we have

(7) $$e^{-Qt}\| L_t\phi\| \le \| P_t(1_D\phi)\| \le (2\pi t)^{-d/2}\int_D \phi\, dm.$$

Using (7) with t = 1 (say) and combining it with (6), we obtain

(8) $$\lim_{t\to\infty} \| L_t\, 1\| = 0.$$

Hence there exists s such that $\| L_s\, 1\| = \eta < 1$. Since

(9) $$\| L_{s+t}\, 1\| \le \| L_s\, 1\| \| L_t\, 1\|$$

for all $s \ge 0$, $t \ge 0$, and $\| L_s\, 1\| \le e^{Qs}$ trivially, it follows that

(10) $$\| L_t\, 1\| \le e^{Qs}\, \eta^{[t/s]},$$

which is equivalent to (4). □

From now on we shall assume that the gauge for (D,q) is finite. If f and therefore h is bounded, the following proposition is contained in Theorem 3.2(b) of [3]. For unbounded but integrable f the proof relies on the inequality (2), which will be assumed throughout the rest of the paper.

PROPOSITION 2. *For each* x *in* D, *we have*

$$\int_0^\infty L_t h(x)dt < \infty. \tag{11}$$

PROOF. Since $h \geq 0$ and h is harmonic in D, h is excessive with respect to (P_t^D) (see [1], §4.5, Theorem 3). Hence for $t \geq 0$,

$$L_t h \leq e^{Qt}\, P_t^D h \leq e^{Qt}\, h. \tag{12}$$

Next, since $h \in L^1(D,m)$, we have by (7)

$$\| L_1 h\| \leq e^Q (2\pi)^{-d/2} \int_D h\, dm < \infty, \tag{13}$$

and consequently by (4)

$$\| L_{t+1}\, h\| \leq \| L_1 h\| \| L_t\, 1\| \leq \| L_1 h\| C_2\, e^{-C_1 t}. \tag{14}$$

It follows from (12) and (14) that

$$\int_0^\infty L_t h\, dt \leq e^Q\, h + \| L_1 h\| \int_0^\infty C_2\, e^{-C_1 t}\, dt < \infty. \qquad \square$$

THEOREM 3. *If* $u(f;\cdot) \not\equiv \infty$ *in* D, *then we have in* D

$$u(f;\cdot) = h(f;\cdot) + \int_0^\infty L_t(qh)dt. \tag{15}$$

PROOF. Since q is bounded, Proposition 2 implies that

$$\int_0^\infty L_t(|q|h)(x)dt < \infty$$

for each x in D. This permits the application of Fubini's theorem in the calculation below, where $\tau = \tau_D$:

$$\int_0^\infty L_t(qh)dt = \int_0^\infty E^x\{t < \tau;\ e(t)q(X_t)E^{X(t)}[f(X(\tau))]\}dt$$

$$= \int_0^\infty E^x\{1_{\{t<\tau\}}\ e(t)q(X_t)f(X(\tau))\}dt$$

$$= E^x\{\int_0^\tau e(t)q(X_t)dt\cdot f(X(\tau))\}$$

$$= E^x\{[e(\tau) - e(0)]f(X(\tau))\} = u(f;x) - h(f;x). \qquad \square$$

THEOREM 4. *There exists* C_3 *such that for all positive* $f \in L^1(\partial D,\sigma)$ *we have*

$$\int_D u(f;x)m(dx) \le C_3 \int_{\partial D} f(y)\sigma(dy). \tag{16}$$

PROOF. We have by (12)

$$\int_0^1 L_t(qh)dt \le Q\,e^Q\,h;$$

and by (13) and (14)

$$\int_1^\infty L_t(qh)dt \le Q\|L_1h\|C_2C_1^{-1} \le Q\,e^Q(2\pi)^{-d/2}\ C_2C_1^{-1}\int h\,dm.$$

Hence we may integrate (15) over D to obtain (16) with

$$C_3 = C + Q\,e^Q\,C + Q\,e^Q(2\pi)^{-d/2}\ C_2C_1^{-1}Cm(D). \qquad \square$$

An interesting application is to the boundary value problem for the

Schrödinger equation. We shall generalize Theorem 1.3 of [3] to the case where $f \in L^1(\partial D, \sigma)$. The following lemma furnishes the clue.

LEMMA 5. *Let* $z \in \partial D$, $B(z,r)$ *be the ball with center* z *and radius* $r > 0$, *and* $\tau_r = \tau_{B(z,r)}$ *for short. Put also* $\tilde{u} = 1_D u$, *and define for* $x \in B(z,r)$

$$w_r(x) = E^x\{e(\tau_r)\tilde{u}(X(\tau_r))\}. \tag{17}$$

Then there exists $r_0 > 0$ *such that for* (m) *a.e.* $r \in (0, r_0]$, *we have*

$$\sup_{x \in \overline{B(z,r)}} w_r(x) < \infty. \tag{18}$$

PROOF. As a consequence of Theorem 4, we have for every $r > 0$

$$\int_{B(z,r)} \tilde{u}\, dm \le \int_D u\, dm < \infty. \tag{19}$$

Hence by Fubini's theorem, for (m) a.e. r we have

$$\int_{\partial B(z,r)} \tilde{u}\, d\sigma < \infty \tag{20}$$

where of course σ denotes the area on the sphere. (I am indebted to K. M. Rao for this brilliant stroke.) It is well known that τ_r and $X(\tau_r)$ are stochastically independent under P^z. Hence if we take x to be z in (17), we have

$$w_r(z) \le E^z\{e^{Q\tau_r}\, \tilde{u}(X(\tau_r))\} = E^z\{e^{Q\tau_r}\}E^z\{\tilde{u}(X(\tau_r))\} \tag{21}$$

$$= E^z\{e^{Q\tau_r}\} \frac{1}{\sigma(\partial B(z,r))} \int_{\partial B(z,r)} \tilde{u}\, d\sigma.$$

It is well known that there exists $r_0 > 0$ such that $E^z\{e^{Q\tau_r}\} < \infty$ for $r \le r_0$. Hence for such values of r we have $w_r(z) < \infty$ by (20) and (21). But the finiteness of w_r at any point in $B(z,r)$ implies its boundedness in $\overline{B(z,r)}$ by Theorem 1.2 of [3], which was recalled earlier; thus (18) follows. □

THEOREM 6. *Suppose* $z \in \partial D$ *and* z *is regular for* D^c. *Let* $f \in L^1(\partial D,\sigma)$ *and* f *be continuous at* z. *Then we have*

$$\lim_{\bar{D} \ni x \to z} u(f;x) = f(z). \tag{22}$$

PROOF. We may suppose $f \ge 0$. There exists $r_1 > 0$ such that f is bounded in $\partial D \cap B(z,r_1)$, say by M. We may suppose this r_1 to be less than the r_0 in Lemma 5. Let $r < r_1$ and write B for $B(z,r)$. We have, by the strong Markov property, for $x \in B \cap D$,

$$u(x) = E^x\{\tau_B < \tau_D;\ e(\tau_B)u(X(\tau_B))\} + E^x\{\tau_D \le \tau_B;\ e(\tau_D)f(X(\tau_D))\}. \tag{23}$$

The second expectation above is bounded by

$$M_1 = \sup_{x \in B} E^x\{e^{Q\tau_B}\}M < \infty,$$

because r_0 was so chosen that the supremum above is finite. The first expectation in (23) does not exceed the $w_r(x)$ in (17), hence bounded in $B(z,r)$ for (m) a.e. $r < r_1$ by Lemma 5. The upshot is that u is bounded in some ball $B(z,r_2)$, and that is sufficient to prove (22) by the same method as used in the case where f is bounded. Specifically, the only difference in the present more general case is the estimation of the quantity

$$E^x\{T_r < \tau_D;\ e(T_r)u(X(T_r))\} \tag{24}$$

(denoted by $u_1(x)$ in the proof of Theorem 1.3 in [3]), where

$$T_r = \inf\{t > 0 \,|\, |X(t) - X(0)| \geq r\}.$$

If $x \in B(z,r_2)$ and $r < r_2/2$, then under P^x we have $X(T_r) \in B(z,r_2)$ and so $u(X(T_r)) \leq M_2$ (say). Therefore the quantity in (24) does not exceed

$$E^x\{T_r < \tau_D;\ e^{QT_r}\}M_2,$$

which converges to zero as $x \to z$, as shown in [3]. The rest of the proof is exactly the same as given there. □

When $q \equiv 0$, $u(f;x) \equiv h(f;x)$ defined in (1), and Theorem 6 reduces to the classical Dirichlet boundary value problem for a harmonic function. The case where the boundary function f is unbounded does not seem well known, but Jang-Mei G. Wu sent me a proof where D is assumed only to be a Lipschitz domain. Her proof relies on a boundary Harnack principle (see [4]) and is quite different from the above. The result in the Schrödinger case may be new.

Let us now consider the validity of the fundamental inequality (2). If D is the unit ball $B = B(0,1)$ in R^3, the Poisson formula gives

$$h(f;x) = \frac{1}{4\pi}\int_{\partial B} \frac{1-|x|^2}{|x-y|^3} f(y)\sigma(dy). \tag{25}$$

Putting $f \equiv 1$ in the above and then integrating with respect to x we obtain

$$\frac{4\pi}{3} = \frac{1}{4\pi} \int_{\partial B} \Big[\int_B \frac{1 - |x|^2}{|x - y|^3} m(dx)\Big] \sigma(dy).$$

But the inner integral is independent of y by spherical symmetry, hence its value is $4\pi/3$. It follows that

$$\int_B h(f;x)m(dx) = \frac{1}{3} \int_{\partial B} f(y)\sigma(dy); \tag{26}$$

namely (2) is an equality with $C = 1/3$. Suppose next $\pi_D(x,y)\sigma(dy)$ is the generalized Poisson kernel for D, so that

$$h(f;x) = \int_{\partial D} \pi_D(x,y)f(y)\sigma(dy), \quad x \in D. \tag{27}$$

Then (2) will follow provided the function

$$y \to \int_D m(dx)\pi_D(x,y)$$

is bounded on ∂D. If $g_D(x,y)$ is the Green's function for D, then $\pi_D(x,\cdot)$ is the normal derivative of $g_D(x,\cdot)$ at ∂D, when the latter is smooth enough. This will become infinite at ∂D, but it is plausible that its integral over D may be bounded. For instance, suppose there is a function ϕ from R^1 to R^1 satisfying $\int_0^R \phi(r)r^{d-1}\,dr < \infty$ for every $R < \infty$, such that

$$\pi_D(x,y) \le \phi(|x - y|), \quad x \in D, \quad y \in \partial D; \tag{28}$$

then the result is true. For the ball above $\phi(r) = r^{-2}$. On the basis of such considerations the inequality was surmised to hold under fairly general conditions. A rigorous proof may be based on Green's formula, as suggested by Rao, provided that the existence and continuity of the normal derivative of h at the boundary, as well as that of $\int_D g_D(\cdot,y)m(dy)$, is assumed. The proviso will be in force if ∂D belongs to the class $C^{2,\alpha}$ by Schauder's results, and more broadly, if ∂D belongs to the class $C^{1,\alpha}$ by Widman's results. The reader is referred to standard treatises on these results for information. Without using them, and when ∂D belongs to the class C^2, Falkner has given a direct geometrical proof of (2) when σ is replaced by the harmonic measure $H(x_0,\cdot)$ for some $x_0 \in D$. He also gave an example in which (2) is false. It is known that (2) may be false for a bounded Lipschitz domain.

Last but not least, for a class of unbounded functions, q, a local form of Theorem 4 was given in [6]; namely for each x there exists a sufficiently small ball $B(x,r)$ for which (16) holds when D is $B(x,r)$. Indeed, then the left member of (16) may be replaced by $u(f;x)$, by an easy application of Harnack's inequality.

N.B. I learned belatedly that an inequality by Widman ([7], Theorems 2.3 and 2.4) yields (28) with $\phi(r) = \text{const. } r^{1-d}$, provided ∂D belongs to $C^{1,\alpha}$.

References

[1] K. L. CHUNG. *Lectures from Markov Processes to Brownian Motion.* Grundlehren 249, Springer-Verlag, Berlin, 1982.

[2] K. L. CHUNG and P. LI. Comparison of probability and eigenvalue methods for the Schrödinger equation. *Advances in Math.* To appear.

[3] K. L. CHUNG and K. M. RAO. Feynman-Kac functional and the Schrödinger equation. *Seminar on Stochastic Processes, 1981,* pp. 1-29. Birkhäuser, Boston, 1981.

[4] J-M. G. WU. Comparisons of kernel functions, boundary Harnack principle and relative Fatou theorem on Lipschitz domains. *Ann. Inst. Fourier 28, 4* (1978), 147-167.

[5] ZHAO ZHONG-XIN. Local Feynman-Kac semigroup. To appear.

[6] M. AIZENMAN and B. SIMON. Brownian motion and Harnack inequality for Schrödinger operators. *Comm. Pure Appl. Math.* To appear.

[7] K.-O. WIDMAN. Inequalities for the Green function and boundary continuity of the gradient of solutions of elliptic differential equations. *Math. Scand. 21* (1967), 17-37.

K. L. CHUNG
Department of Mathematics
Stanford University
Stanford, CA 94305

Seminar on Stochastic Processes, 1982
Birkhäuser, Boston, 1983

REGENERATIVE SYSTEMS AND MARKOV ADDITIVE PROCESSES

by

E. ÇINLAR* and H. KASPI

Consider a regenerative system $(\Omega, F, F_t, \theta_t, X_t, P_x; M)$ in the sense of Maisonneuve [6] and Jacod [4]; here M is the regeneration set and X is the "mark" process. When the mark process is trivial, that is, when $X_t(\omega) = x_0$ for all $t \in M(\omega)$ for some fixed point x_0, Maisonneuve [7] showed that M is the image of an increasing Lévy process. When M has no isolated points and every stopping time T with $[T] \subset \bar{M}\backslash M$ is totally inaccessible, Jacod showed that (X,M) is the image of a Markov additive process; see [4] and also [6]. Our aim is to extend the latter result by allowing M to have isolated points and limits of isolated points.

Section 1 is devoted to notations and preliminaries regarding regenerative systems; principles are adopted from Blumenthal and Getoor [1] and Maisonneuve [6]. In Section 2 we construct a continuous additive functional that increases on the perfect part of M. In Section 3 we construct a pure jump additive "functional" which jumps at the isolated points of M by suitably chosen exponential amounts independent of (X,M). Then, if (τ_s) is the right continuous inverse of the sum of the two additive functionals constructed, the process (X_{τ_s}, τ_s) is

---

*Research supported by AFOSR Grant No. 80-0252.

a Markov additive process whose "image" is the regenerative system we started with; details of this are put in Section 4.

The assumption that stopping times whose graphs are contained in $\bar{M}\backslash M$ are totally inaccessible is fairly restrictive; but, in many interesting cases (such as visits of a Hunt process to a Borel set), this assumption can be dropped (see Kaspi [5]).

1. Definitions and Preliminaries

Let (Ω, F^o, P) be a probability space, (F_t^o) a right continuous filtration with $\vee_t F_t^o = F^o$, (θ_t) a family of shift operators on Ω, and G a sub-σ-algebra of F^o. We assume that the shifts θ_t and the mapping $\theta: (t,\omega) \to \theta_t\omega$ from $\mathbb{R}_+ \times \Omega$ into Ω satisfy

(1.1) $$\theta_t \in G/G, \quad \theta_t \in F_{s+t}^o/F_s^o, \quad \theta \in (R_+ \times F^o)/F^o.$$

Let E be a Borel subset of a compact metric space and E its Borel subsets. For each $\omega \in \Omega$, let $M(\omega)$ be a right-closed subset of $\mathbb{R}_+$, and let $X(\omega)$ be a mapping from $M(\omega)$ into E; we write $X_t(\omega)$ for the image of $t \in M(\omega)$ under $X(\omega)$.

Finally, let $(P_x)_{x \in E}$ be a family of probability measures on (Ω, F^o) such that $x \to P_x(G)$ is E-measurable for every $G \in F^o$. We let F denote the completion of F^o with respect to all $P_\mu = \int \mu(dx)P_x$, μ a finite measure on E, and let F_t be the σ-algebra generated by F_t^o and all sets in F that are P_μ-null for every μ.

(1.2) DEFINITION. The collection (G, M, X, P_x) is said to be a regenerative system over $(\Omega, F, F_t, \theta_t, P)$ provided that the following hold.

(i) $0 \in M(\omega)$ for P^x – almost every ω for all x.

(ii) *Measurability*: X is progressive, that is, for every $B \in \mathcal{E}$, the set $\{(t,\omega): t \in M(\omega),\ X_t(\omega) \in B\}$ is progressive relative to $(\mathcal{F}_t)$.

(iii) *Homogeneity*: For every $\omega \in \Omega$ and $t \in M(\omega)$,

$$M(\theta_t\omega) \cap (0,\infty) = (M(\omega) - t) \cap (0,\infty),$$

$$X_s(\theta_t\omega) = X_{t+s}(\omega) \quad \text{if} \quad s \in M(\theta_t\omega).$$

(iv) *Strong Markov property*: For every $Z \in b\mathcal{G}$ and every stopping time T of $(\mathcal{F}_t)$ such that $T(\omega) \in M(\omega)$ for almost every ω in $\{T < \infty\}$,

$$E[Z \circ \theta_T | \mathcal{F}_T] = E_{X_T}[Z] \quad \text{a.s. on} \quad \{T < \infty\}. \qquad \square$$

We shall always assume that M is a minimal right-closed set (every point of M that is isolated from the right is also isolated from the left) and let $\bar{M}$ denote the closure of M. We shall make two additional regularity assumptions.

HTI: *Every* $(\mathcal{F}_t)$*-stopping time with* $[T] \subset \bar{M}\backslash M$ *is totally inaccessible.*

HRC: X *is right continuous along* M, *that is, for every decreasing sequence* $(t_n) \subset M(\omega)$ *with limit* t, $X_{t_n}(\omega) \to X_t(\omega)$.

Under these assumptions we shall show that there exists a strong Markov additive process $(\hat{\Omega}, M, M_u, \sigma_u, Y_u, \tau_u, \hat{P}_x)$ in the sense of Çinlar [2] such that $\hat{\Omega}$ and $\hat{P}_x$ have the forms $\Omega \times \Omega'$ and $P_x \times P'$ respectively, and for $\hat{P}_x$ - almost every $(\omega,\omega') \in \hat{\Omega}$,

(1.3) $\{(t,x): t \in M(\omega),\ X_t(\omega) = x\} = \{(t,x): \tau_u(\omega,\omega') = t,\ Y_u(\omega,\omega') = x \text{ for some } u\}.$

In looser speech, we refer to (1.3) by saying that the regenerative system (M,X) is the image of the Markov additive process (Y,τ).

Returning to the regenerative system, we introduce the following much as in [6]. We let M_i denote the set of isolated points of M, M_i^g the left-closure of M_i, and M_i^d the right-closure of M_i. Let

$$(1.4)\qquad R = \inf\{t > 0: t \in M\}, \quad S = \inf\{t \geq 0: t \in M_i\},$$

$$(1.5)\qquad G_t = \sup\{s \leq t: s \in \bar{M}\}, \quad D_t = \inf\{s > t: s \in M\}, \quad R_t = D_t - t$$

(with the usual conventions that $\inf \emptyset = \infty$, $\sup \emptyset = 0$). We suppose that there is a cemetery point ω_Δ in Ω for which $M(\omega_\Delta) = \emptyset$. We extend the definition of $X(\omega)$ from $M(\omega)$ onto $\overline{\mathbb{R}}_+$ by setting $X_t(\omega) = \Delta$ for all $t \notin M(\omega)$ and $t = \infty$, and assume that $\theta_\infty\omega = \omega_\Delta$ for all ω.

Define $\mathcal{H}_t$ to be the completion of

$$(1.6)\qquad \mathcal{H}_t^o = \sigma(X_0) \vee \sigma(R_s, X_{D_s};\ s \leq t)$$

with respect to all P_μ, μ a finite measure, and let $\mathcal{H}^o = \mathcal{H}_\infty^o$ and $\mathcal{H} = \mathcal{H}_\infty$. Note that

$$\mathcal{F} \supset \mathcal{G} \supset \mathcal{H}, \qquad \mathcal{F}_{D_t} \supset \mathcal{H}_t,$$

where $\mathcal{F}_{D_t}$ makes sense since D_t is a stopping time of $(\mathcal{F}_t)$ because of progressiveness of M (see Definition (1.2ii)). The following two results are mostly in Jacod [4] and mostly easy. We list them without proofs.

(1.7) LEMMA. *Each* G_t *is* $\mathcal{F}_t$*-measurable and is an* $(\mathcal{H}_t)$*-stopping time;* S *is both an* $(\mathcal{F}_t)$*-stopping time and an* $(\mathcal{H}_t)$*-stopping time.*

(1.8) LEMMA. *There exist three sequences* (S_n), (T_n), *and* (U_n) *of* $(\mathcal{F}_t)$*-stopping times with disjoint graphs such that*

$$(1.9) \qquad [S_n] \subset M_i, \quad [T_n] \subset M_i^d \backslash M_i, \quad [U_n] \subset M \backslash M_i^d,$$

$$(1.10) \qquad \bigcup_n ([S_n, D_{S_n}) \cup [T_n] \cup [U_n, U_n + S \circ \theta_{U_n}]) = \mathbb{R}_+ \times \Omega. \qquad \square$$

In [4] the sequences (T_n) and (U_n) were chosen to cover the sets of right-end-points of the intervals contiguous to $\overline{M \backslash M_i}$ and $M_i^g \backslash M_i$ respectively. By that choice, the stochastic intervals $[U_n, U_n + S \circ \theta_{U_n})$ are not necessarily disjoint. For our purposes we need the following strengthening.

(1.11) LEMMA. *The* U_n *can be so chosen that the intervals* $[U_n, U_n + S \circ \theta_{U_n})$ *are disjoint.*

PROOF. Roughly speaking, if two such intervals intersect, then one is contained in the other and can be deleted. Throughout this proof, we write Θ_n for θ_{U_n}.

a) By the homogeneity condition on M, there exists a set $A \in \mathcal{G}$ with $P_x(A) = 0$ for all x such that, for $\omega \in A^c$ and $t \in M(\omega)$, we have $M_i(\theta_t \omega) \cap (0,\infty) = (M_i(\omega) - t) \cap (0,\infty)$. Thus,

$$S(\theta_t \omega) = S(\omega) - t \quad \text{if} \quad \omega \in A^c \cap \{S > t, \ t \in M\}.$$

Let

$$A_n = \{\omega\colon S(\Theta_n \omega) > t \ \text{ and } S(\Theta_n \omega) - t \neq S(\theta_t \circ \Theta_n(\omega)) \ \text{ for some } t \in M(\Theta_n \omega)\}.$$

Then, since $[U_n] \subset M$, $A \in \mathcal{G}$, and $A_n \subset \Theta_n^{-1} A$, the strong Markov property (1.2iv) yields

$$P_x(A_n) \le P_x(\Theta_n^{-1}A) = E_x[P_{X(U_n)}(A)] = 0, \quad x \in E.$$

Let

(1.12) $$B = \bigcup_n A_n.$$

Then, $P_x(B) = 0$ for all x, and for $\omega \notin B$, for all n and k,

$$U_n(\omega) < U_k(\omega) < U_n(\omega) + S(\Theta_n\omega) \Rightarrow U_k(\omega) + S(\Theta_k\omega) = U_n(\omega) + S(\Theta_n\omega).$$

In other words, for $\omega \notin B$, if two intervals (of the kind we are considering) intersect, then one interval is contained in the other and their right-end-points coincide.

b) Fix n and let

(1.13) $$B_n = \Omega \setminus \bigcup_m \{U_m < U_n, U_m + S \circ \Theta_m > U_n\}.$$

Since U_m, U_n, $U_m + S \circ \Theta_m$ are all (F_t)-stopping times, $B_n \in F_{U_n}$ and

(1.14) $$V_n = \begin{cases} U_n & \text{on } B_n \\ \infty & \text{on } \Omega \setminus B_n \end{cases}$$

defines an (F_t)-stopping time.

c) We claim that the U_n in Lemma (1.8) can be replaced by the V_n and then the conclusion of Lemma (1.11) holds. It is easy to show that, for $\omega \notin B$, the intervals $[V_n(\omega), V_n(\omega) + S \circ \theta_{V_n}(\omega))$ are disjoint. The remaining claim is that (1.10) remains true when the U_n are replaced by the V_n. To show this, it is sufficient to show that

(1.15) $$\bigcup_n [U_n, U_n + S \circ \Theta_n) \doteq \bigcup_n [V_n, V_n + S \circ \Theta_{V_n}).$$

d) Fix $\omega \in \Omega \backslash B$, write u_n for $U_n(\omega)$ and v_n for $U_n(\omega) + S \circ \theta_n(\omega)$. Suppose $t \in [u_k, v_k)$ for some k. To show (1.15), we need to show that there exists n such that

$$t \in [V_n(\omega), V_n(\omega) + S \circ \theta_{V_n}(\omega)). \tag{1.16}$$

For this purpose, define

$$N = \{n: v_n = v_k\}, \quad u = \inf_{n \in N} u_n. \tag{1.17}$$

If N is finite, then $u = u_n$ for some $n \in N$ and (1.16) holds for that n. Suppose N is infinite and suppose u is not equal to any of the u_n. Since the graphs of the U_n are disjoint, u must be the limit from right of points that are in $M(\omega)$. Thus, $u \in M(\omega) \backslash M_i(\omega)$. Further, u cannot be in $M_i^d(\omega)$, because none of the intervals (u_n, v_k) contains points in $M_i(\omega)$, and, therefore, neither does their union (u, v_k). Hence, $u \in M(\omega) \backslash M_i^d(\omega) \subset \bigcup_n [u_n, v_n)$, and there is m such that $u \in [u_m, v_m)$. Consider this particular m. If $u_m < u$, by the definition (1.17) of u, m must not belong to N. Therefore, since ω is not in B, $[u_m, v_m)$ does not intersect any $[u_n, v_n)$ with $n \in N$. But $u \in [u_m, v_m)$ and $[u_m, v_m)$ is half open, and this contradicts the definition of u.

Hence, $u \in \bigcup_{n \in N} [u_n, v_n)$, and by (1.17), there is $n \in N$ such that $u = u_n$ and $t \in [u_n, v_n)$. But, for this n, $V_n(\omega) = u_n$ because $u_m < u_n$ implies $v_m < u_n$ by part (a) of the proof. Now (1.16) holds for this n.

2. The Perfect Part of the Regeneration Set

In this section we shall construct a continuous additive functional that increases on the perfect part of M. The construction relies heavily on the results of [4] and [6].

As in [4] we assume that to each $x \in E$ there corresponds $\bar{\omega}_x \in \Omega$ for which $M(\bar{\omega}_x) = \{0\}$ and $x_0(\bar{\omega}_x) = x$. This assumption is not very serious. The following is the main result of this section.

MAIN THEOREM. *There exists a process* (L_t^c) *that is continuous, R-additive, and adapted to* (F_t); *its 1-potential is bounded by* 1; *and its set of times of right increase is the right-closure of* $M \backslash M_i$.

The remainder of this section is devoted to the construction of L^c and the proof of this theorem.

Recall (1.4) that S is the first isolated point of M. Define

$$(2.1)\qquad M' = M \cup [S,\infty); \qquad D = \{x \in E: P_x(S > 0) = 1\},$$

$$X_t'(\omega) = X_{t \wedge S(\omega)}(\omega), \quad t \in M(\omega);$$

$$\theta_t'\omega = \theta_t\omega \quad \text{if } t < S(\omega), \qquad \theta_t'\omega = \bar{\omega}_{X_S(\omega)} \quad \text{if } t \geq S(\omega);$$

$$P_x' = P_x \quad \text{if } x \in D, \qquad P_x' = \delta_{\bar{\omega}_x} \quad \text{if } x \in E \backslash D.$$

Define R_t', D_t', etc. from M' as R_t, D_t, etc. are defined from M; put

$$(2.2)\qquad K_t^o = \sigma(R_s', X_{D_s'}'; s \leq t), \qquad K^o = \vee_t K_t^o,$$

and let K_t and K be the usual right continuous completions of K_t^o and K^o with respect to all P_μ', μ a finite measure on E.

The new part of the following was proved in [4]; see (1.6)ff. for the rest.

(2.3) LEMMA. *We have* $H^{o}_{t+} \supset K^{o}_{t+}$ *and* $F_{D_t} \supset H_t \supset K_t$.

(2.4) THEOREM. *The collection* (G, M', X', P'_x) *is a regenerative system over* $(\Omega, F, F_t, \theta'_t, P)$ *satisfying* HTI *and* HRC. *Moreover,* M' *has no isolated points.*

Proof of this is essentially contained in [4]; we merely add that the required measurability properties follow from the facts that S is G-measurable, S is an (F_t)-stopping time, and the measurability assumptions for the original system.

Now, the results of [6] apply to the perfect regenerative system (M',X'), and there exists a perfect continuous local time (L'_t) of equilibrium of order 1 with the following properties (in fact, Maisonneuve works with the canonical realizations, but we can use the mapping $\pi: \omega \to (R'_t(\omega), X'_{D'_t}(\omega))_{t \geq 0}$ to go to the canonical space and define $L'_t(\omega) = L_t(\pi\omega)$ where L is his local time): except on a set of P'_x measure 0 for all x,

$$(2.5) \qquad L'_{D'_s} = L'_s, \quad L'_{D'_s + t} = L'_{D'_s} + L'_t \circ \theta'_{D'_s}, \qquad s,t \geq 0;$$

(L'_t) is adapted to (K_t); and almost surely (P'_x for all x), the points of increase of (L'_t) are exactly the points of M'.

Recall the stopping time U_n described in Lemma (1.8), and assume that they are chosen (as they can be, see Lemma (1.11)) so that the intervals $[U_n, U_n + S \circ \theta_{U_n})$ are disjoint. We now introduce the local time L^c of the perfect part of M:

$$(2.6) \qquad L^n_t(\omega) = L'_t \circ \theta_{U_n}(\omega),$$

$$(2.7)\qquad L_t^c = \sum_n (L^n_{t-U_n}\, I_{\{U_n \le t \le U_n + S\circ\theta_{U_n}\}} + L^n_S\, I_{\{U_n + S\circ\theta_{U_n} < t\}}).$$

We introduce the following for reasons of typographical ease in this section:

$$(2.8)\qquad X_t^+ = X_{D_t},\quad F_t^+ = F_{D_t},\quad \Theta_n = \theta_{U_n},\quad V_n = U_n + S\circ\Theta_n.$$

(2.9) LEMMA. L_t^n *is* $F^+_{U_n+t}$ *measurable.*

PROOF. By the definition (2.6) of L^n, $\{L_t^n \le s\} = \Theta_n^{-1}\{L_t^! \le s\}$; hence, it is sufficient to show that

$$\Theta_n^{-1}\, K_t \subset F^+_{U_n+t}.$$

Since U_n is a stopping time of (F_t) and therefore of (F_t^+), $U_n + s$ is a stopping time of (F_t^+). Since (R_t, X_t^+) is right continuous and therefore progressive, $(R_{U_n+s}, X^+_{U_n+s})$ is $F^+_{U_n+s}$ measurable. But, since $[U_n] \subset M\backslash M_i^d$, almost surely,

$$R_{U_n+s} = R_s\circ\Theta_n,\quad X^+_{U_n+s} = X_{D(U_n+s)} = X_{D_s}\circ\Theta_n = X_s^+\circ\Theta_n.$$

It follows that $(R_s\circ\Theta_n, X_s^+\circ\Theta_n)$ is $F^+_{U_n+s}$ measurable for all s. So, $\Theta_n^{-1}(K^o_{t+}) \subset F^+_{U_n+t}$. Hence, to complete the proof, it is sufficient to show that

$$(2.10)\qquad A\in K_t,\quad P_\mu^!(A) = 0 \text{ for all } \mu \Rightarrow \Theta_n^{-1}(A)\in F^+_{U_n+t}.$$

Let A be such a set, fix a finite measure μ, and let ν be the distribution of $X(U_n)$ under P_μ. There exists a set $B\in K^o$ such that $B\supset A$ and $P_\nu^!(B) = 0$ (since $P_\nu^!(A) = 0$ for all ν). Then,

$$P_\mu(\Theta_n^{-1}(B)) = E_\mu[P_{X(U_n)}(B)] = E_\mu[P'_{X(U_n)}(B)] = P'_\nu(B) = 0,$$

where the second equality is justified by noting the definition (2.1) for P'_x and by noting that $X(U_n) \in D$ almost surely on $\{U_n < \infty\}$. Thus, for each μ, there is $B \in \mathcal{K}^\circ \subset \mathcal{H}^\circ \subset \mathcal{F}$ such that $\Theta_n^{-1}(A) \subset \Theta_n^{-1}(B)$ and $P_\mu(\Theta_n^{-1}(B)) = 0$. Since each $\mathcal{F}_s^+$ contains all the null sets of $\mathcal{F}$, this shows that $\Theta_n^{-1}(A) \in \mathcal{F}_s^+$ for all s. This in turn implies that $\Theta_n^{-1}(A) \cap \{U_n + t \le s\} \in \mathcal{F}_s^+$ for all s since $U_n + t$ is a stopping time of $(\mathcal{F}_s^+)$. So, $\Theta_n^{-1}(A) \in \mathcal{F}_{U_n+t}^+$ as desired in (2.10).

(2.11) LEMMA. *For each* n, s, *and* t, $\{U_n \le t, L^n_{t-U_n} \le s\} \in \mathcal{F}_t^+$.

PROOF. Fix n, s, t; let A be the event in question. Then,

$$A = \lim_m \bigcup_{t \ge k/2^m} \{(k-1)/2^m < U_n \le k/2^m, L^n_{t-k/2^m} \le s\}$$

$$= \lim_m \bigcup_{t > k/2^m} \{t \;\; 1/2^m < U_n + t - k/2^m \le t, L^n_{t-k/2^m} \le s\},$$

by the continuity of (L_t^n). By Lemma (2.9),

$$\{L^n_{t-k/2^m} \le s\} \in \mathcal{F}^+_{U_n + t - k/2^m},$$

and $U_n + t - k/2^m$ is a stopping time of $(\mathcal{F}_t^+)$ for all k and m such that $t \ge k/2^m$. Hence, each event in the union is $\mathcal{F}_t^+$-measurable, and therefore, so is the union and the limit.

(2.12) PROPOSITION. L^c *defined by* (2.7) *is finite valued, continuous, and adapted to* $(\mathcal{F}_t)$.

PROOF. We shall show that L^c is adapted to $(\mathcal{F}_t^+)$ and is finite valued. Then, the monotone convergence theorem implies the continuity of L^c since each L^n is an increasing continuous process; and the continuity of L^c together with its being adapted (F_t^+) implies that L^c is adapted to (F_t) as well (see [6]).

a) Recall the notations (2.8). The functional

$$(2.13) \qquad \hat{L}_t^n = L_{t-U_n}^n \, I_{\{U_n \le t\}} - (t - V_n) \, I_{\{V_n \le t\}}, \qquad t \ge 0,$$

is (F_t^+) adapted by Lemma (2.11) and by the fact that V_n is a stopping time of (F_t^+). Hence, $\sum_n \hat{L}^n = L^c$ is adapted to (F_t^+).

b) We now show that $L_t^c < \infty$. Since the U_n are (F_t^+) stopping times and L' is adapted to (F_t^+),

$$(2.14) \qquad E_x \int_0^\infty e^{-t} \, dL_t^c = E_x \sum_n \int_{[U_n, V_n]} e^{-t} \, dL_t'(\theta_n)$$

$$= E_x \sum_n e^{-U_n} E_{X(U_n)} \int_0^S e^{-t} \, dL_t'.$$

If $x \in D$, then

$$E_x \int_0^S e^{-t} \, dL_t' = E_x \int_0^\infty e^{-t} \, dL_t' - E_x[e^{-S} E_{X(S)} \int_0^\infty e^{-t} \, dL_t']$$

$$= 1 - E_x[e^{-S}] = E_x[1 - e^{-S}].$$

Since $X(U_n) \in D$ almost surely on $\{U_n < \infty\}$, putting this into (2.14),

$$(2.15) \qquad E_x \int_0^\infty e^{-t} \, dL_t^c = E_x \sum_n e^{-U_n} E_{X(U_n)}[1 - e^{-S}]$$

$$= E_x \sum_n (e^{-U_n} - e^{-V_n}) \le E_x \int_0^\infty e^{-t} \, dt = 1$$

since the intervals $[U_n, V_n)$ are disjoint by the way the U_n are selected.

(2.16) PROPOSITION. L^c *is continuous,* R-*additive, adapted to* (F_t), *and its* 1-*potential is bounded by* 1.

PROOF. In view of the preceding proposition and (2.15), we need to show only R-additivity. Since $\hat{L}_n$ defined by (2.13) does not increase outside (U_n, V_n), and because of (2.5), we have

$$L^c_{D_t} = L^c_t, \quad t \geq 0.$$

Thus, by (2.7) and (2.8), further writing $\theta^+_t = \theta_{D_t}$,

$$\begin{aligned} L^c_{D_t+s} &= \sum_n L^n_{D_t+s-U_n} I_{\{U_n \leq D_t+s \leq V_n\}} + \sum_n L^n_S I_{\{V_n < D_t+s\}} \\ &= \sum_n L^n_{D_t-U_n} I_{\{U_n \leq D_t \leq V_n\}} + \sum_n L^n_S I_{\{V_n < D_t\}} \\ &\quad + \sum_n L^n_{s-U_n \circ \theta^+_t}(\theta^+_t) I_{\{U_n \circ \theta^+_t \leq s \leq V_n \circ \theta^+_t\}} + \sum_n L^n_S \circ \theta^+_t I_{\{V_n \circ \theta^+_t < s\}} \\ &= L^c_{D_t} + \Big(\sum_n L^n_{s-U_n} I_{\{U_n \leq s \leq V_n\}} + \sum_n L^n_S I_{\{V_n < s\}}\Big) \circ \theta^+_t \\ &= L^c_{D_t} + L^c_s \circ \theta_{D_t}, \end{aligned}$$

where we used the fact that $[D_t] \subset M$ and that M_0 is homogeneous, where M_0 is the set consisting of the points of $M^g_i \backslash M_i$ and the right-end-points of intervals contiguous to $\overline{M \backslash M_i}$. □

In view of the preceding proposition, the following completes the proof of Main Theorem of this section.

(2.17) PROPOSITION. *The set of points of right increase of* L^c *is the right closure of* $M \backslash M_i$.

PROOF. It is sufficient to show that, for each n, L^n increases on $M' \circ \Theta_n$ almost surely (recall that $\Theta_n = \theta_{U_n}$). But, by the definition (2.6) of L^n, this is the same as saying that $L' \circ \Theta_n$ increases on $M' \circ \Theta_n$ almost surely, or equivalently, for all x, $P_x(\Theta_n^{-1}(A)) = 1$ where A is the event that L' increases on M'. Now, $A \in \mathcal{K} \subset \mathcal{G}$ and $[U_n] \subset M$ and $X(U_n) \in D$ almost surely. So, by the strong Markov property at U_n,

$$P_x(\Theta_n^{-1}(A)) = E_x[P_{X(U_n)}(A)] = 1,$$

since $P_y(A) = P'_y(A)$ for $y \in D$ and $P'_y(A) = 1$ by the way L' is chosen. □

3. Isolated Points of M

To deal with the isolated points and with right limits of such points, we shall construct a "local time" that jumps by independent exponential amounts at each visit to M_i. Further, these exponential amounts will be independent of $\mathcal{F}$. This requires enlarging the original probability space in such a way that the regeneration property is preserved. A construction similar to the one presented here was done in [3] to deal with the jump structure of Hunt processes.

Let η be the exponential distribution on $(\mathbb{R}_+, \mathcal{R}_+)$ with parameter 1, and let $\mathbb{N} = \{0, 1, 2, \ldots\}$, $\mathbb{N}_0 = \{1, 2, \ldots\}$. Let

$$(3.1) \qquad (\overset{\shortmid}{\Omega}, \overset{\shortmid}{\mathcal{F}}, \overset{\shortmid}{P}) = (\mathbb{R}_+, \mathcal{R}_+, \eta)^{\mathbb{N}_0 \times \mathbb{N}_0},$$

and let W_{mn} be the coordinate variables on it. Define

$$(3.2) \qquad (\hat{\Omega}, \hat{\mathcal{F}}, \hat{P}_x) = (\Omega, \mathcal{F}, P_x) \times (\overset{\shortmid}{\Omega}, \overset{\shortmid}{\mathcal{F}}, \overset{\shortmid}{P}),$$

and extend the definitions of X, M, W_{mn} onto $\hat{\Omega}$ in the natural manner (for instance, $M(\omega,\omega') = M(\omega)$ and $W_{mn}(\omega,\omega') = W_{mn}(\omega')$). We define

(3.3) $\lambda(x) = E_x(1 - e^{-R}), \quad x \in E; \quad C = \{x \in E: \lambda(x) > 0\};$

(3.4) $$S_{m1} = \inf\{t \in M_i: \frac{1}{m} < \lambda(X_t) \le \frac{1}{m-1}\};$$

(3.5) $$S_{m,n+1} = D(S_{mn}) + S_{m1} \circ \theta_{D(S_{mn})}, \quad n \ge 1.$$

(3.6) LEMMA. *With* S_n *as in Lemma* (1.8),

$$\bigcup_{m,n} [S_{mn}] = \bigcup_n [S_n] = M_i.$$

PROOF. First note that

$$E_x \sum_{s \in M_i \cap [0,t)} e^{-s}(1 - e^{-R(s)}) \le E_x \int_0^\infty e^{-s}\, ds = 1.$$

On the other hand, the left side is equal to

$$E_x \sum_{s \in M_i \cap [0,t)} e^{-s} E_{X_s}(1 - e^{-R}) \ge e^{-t} E_x \sum_{s \in M_i \cap [0,t)} \lambda(X_s).$$

Therefore, almost surely, there are only a finite number of points s in $M_i \cap [0,t)$ for which $1/m < \lambda(X_s) \le 1/(m-1)$. Hence on $\{S_{mn} < \infty\}$, almost surely,

$$\frac{1}{m} < \lambda(X_{S_{mn}}) \le \frac{1}{m-1}.$$

This completes the proof, since $M_i \subset \{s \in M: X_s \in C\}$. □

Define

(3.7) $$J_{t,m} = \sup\{n: S_{mn} \le t\}, \quad J(t) = (J_{t,m})_{m \in \mathbb{N}_0}.$$

For every sequence $j = (j(m))_{m \in \mathbb{N}_0}$ of integers $j(m) \in \mathbb{N}$, we let $\dot{F}_j$ be the σ-algebra on $\dot{\Omega}$ generated by the coordinate variables W_{mn} with $n \le j(m)$ and $m \in \mathbb{N}_0$, and set

(3.8) $$\hat{F}^o_t = \{A \in F_t \times \dot{F}: I_A(\omega,0) \in \dot{F}_{J(t,\omega)} \text{ for every } \omega \in \Omega\}.$$

We let $\hat{F}_t$ be the right continuous completion (with respect to the $\hat{P}_x$) of $\hat{F}^o_t$ and set $\hat{G} = G \times \dot{F}$. Finally, we define the shifts on $\hat{\Omega}$ by

(3.9) $$\hat{\theta}_t(\omega,\dot{\omega}) = (\theta_t\omega, \dot{\theta}_{J(t,\omega)}\dot{\omega}), \quad (\omega,\dot{\omega}) \in \hat{\Omega},$$

where $\dot{\theta}_j$ is defined for each sequence $j = (j(m))$ so that $W_{mn} \circ \dot{\theta}_j = W_{m,j(m)+n}$ for all m and n in $\mathbb{N}_0$.

(3.10) THEOREM. *The collection* $(\hat{G}, M, X, \hat{P}_x)$ *is a regenerative system over* $(\hat{\Omega}, \hat{F}, \hat{F}_t, \hat{\theta}_t, \hat{P})$ *satisfying* HTI *and* HRC. *Moreover, for every stopping time* T *of* $(\hat{F}_t)$, $R_T > 0$ *almost surely on* $\{T<\infty, X_T \in C\}$.

PROOF is identical to that given in [3, §4.c], replacing stopping times there by regeneration times here. We omit it. We are now ready to define a left-continuous additive "functional" that increases on M^d_i:

(3.11) $$L^d_t = \sum_m \sum_n I_{\{S_{mn} < t\}} \lambda(X_{S_{mn}}) W_{mn}.$$

(3.12) THEOREM. L^d *is left-continuous, adapted to* $(\hat{F}_t)$, *and* R-*additive relative to* $(\hat{\theta}_t)$. *Its* 1-*potential is bounded by* 1, *and its point of right-increase is the right closure of* M_i.

PROOF. We show the claims on additivity and the 1-potential; the others are obvious.

a) Note that the set consisting of the S_{mn} belonging to $[D_t, D_t + s]$ is the same as the set consisting of the $S_{mn} \circ \theta_{D_t}$ belonging to $[0,s)$ shifted to the right by the amount D_t. Hence,

$$L^d_{D_t+s} = \sum_{m,n} I_{\{S_{mn} < D_t\}} \lambda(X_{S_{mn}}) W_{mn}$$

$$+ \sum_{m,n} I_{\{S_{mn} \circ \theta_{D_t} < s\}} \lambda(X_{S_{mn}} \circ \theta_{D_t}) W_{mn} \circ \hat{\theta}_{D_t} = L^d_{D_t} + L^d_s \circ \hat{\theta}_{D_t} .$$

b) Since W_{mn} is independent of (M,X) and has mean 1,

$$\hat{E}_x \int_0^\infty e^{-t} \, dL^d_t = \hat{E}_x \sum_{m,n} W_{mn} \lambda(X_{S_{mn}}) \exp(-S_{mn})$$

$$= E_x \sum_{m,n} \lambda(X_{S_{mn}}) \exp(-S_{mn}).$$

Using the definition (3.3) of $\lambda(x)$, the fact that $[S_{mn}] \subset M$, and the strong Markov property of (M,X) at S_{mn}, we see that the last member is equal to

$$E_x \sum_{m,n} [\exp(-S_{mn}) - \exp(-S_{mn} - R \circ \theta_{S_{mn}})] \le E_x \int_0^\infty e^{-t} \, dt = 1.$$

4. The Markov Additive Process with Range (M,X)

Throughout this section we shall work with $(\hat{\Omega}, \hat{F}, \ldots, \hat{P}_x)$. Recall that all functions defined on Ω or on Ω' are automatically extended onto $\hat{\Omega} = \Omega \times \Omega'$. Putting together the "local times" L^c and L^d constructed in Sections 2 and 3 respectively, we obtain

(4.1) $$L_t = L^c_t + L^d_t, \quad t \ge 0,$$

which is a left-continuous R-additive "functional" such that

$$(4.2) \qquad \{t: L_{t+\varepsilon} > L_t \quad \text{for all} \quad \varepsilon > 0\} = M.$$

Recall that X_∞ was defined to be Δ, a point not in E, and that there is a cemetery point $\hat{\omega}_\Delta$ in $\hat{\Omega}$ for which we have $M(\hat{\omega}_\Delta) = \emptyset$, and $\hat{\theta}_\infty \hat{\omega} = \hat{\omega}_\Delta$ for all $\hat{\omega}$. We define, for all $u \geq 0$,

$$(4.3) \qquad \tau_u = \inf\{t: L_t > u\},$$

$$(4.4) \qquad Y_u = X_{\tau_u}, \quad A_u = u - L_{\tau_u}, \quad \zeta = L_\infty.$$

It is already clear that the range of (τ_u) is M and, more precisely, (1.3) holds. Defining the shifts for (Y,τ) requires delicacy when τ_u belongs to M_i: on $\{\tau_u \in M_i\}$, τ_u coincides with some $S_{jk} = s$, at that point L jumps from L_s to $L_{s+} = L_s + \lambda(X_s)W_{jk}$, and L_{s+} overshoots the level u by the amount $\lambda(X_s)W_{jk} - A_u$; this overshoot has the exponential distribution with mean $\lambda(X_s)$, and is to be equal to $\lambda(X_s)$ times the shifted W_{j1}. Formally, we define the shifts τ_u as follows:

$$(4.5) \qquad \sigma_u(\omega,\omega') = \begin{cases} \hat{\theta}_{\tau_u(\omega,\omega')}(\omega,\omega') & \text{if} \quad \tau_u(\omega,\omega') \notin M_i(\omega), \\ (\theta_{\tau_u(\omega,\omega')}\omega, \sigma'_{u\omega}\omega') & \text{if} \quad \tau_u(\omega,\omega') \in M_i(\omega), \end{cases}$$

where $\sigma'_{u\omega}$ is such that the following holds on $\{\tau_u = S_{jk}\}$;

$$(4.6) \qquad W_{mn} \circ \sigma_u = \begin{cases} W_{mn} \circ \hat{\theta}_{\tau(u)} & \text{if} \quad m \neq j, \quad n \geq 1, \\ W_{jk} - A_u/\lambda(Y_u) & \text{if} \quad m = j, \quad n = 1, \\ W_{j,n-1} \circ \hat{\theta}_{\tau(u)} & \text{if} \quad m = j, \quad n \geq 2. \end{cases}$$

Finally we define

(4.7) $$M = \hat{F}, \quad M^{\circ}_u = \hat{F}_{\tau_{u^-}} \vee \sigma(Y_u),$$

and let M_u be the completion of M°_{u+} in the usual manner.

(4.8) THEOREM. *The process* $(Y,\tau) = (\hat{\Omega}, M, M_u, \sigma_u, Y_u, \tau_u, \hat{P}_x)$ *is a strong Markov additive process, and* (1.3) *holds.*

The remainder of this section is devoted to proving this. There is a slight deviation here from the definition of [2] for Markov additive processes. Here, $\tau_u = \infty$ on $\{u \geq \zeta\}$, whereas in [2], τ_u would be $\tau_{\zeta-}$ on $\{u \geq \zeta\}$.

(4.9) LEMMA. a) *The process* (τ_u) *is increasing, right continuous, and additive relative to the shifts* (σ_u).

b) *For almost every* $(\omega,\omega') \in \hat{\Omega}$, $\{t\colon \tau_u(\omega,\omega') = t \text{ for some } u\} = M(\omega)$.

c) *For every stopping time* T *of* (M_u), τ_T *is a stopping time of* $(\hat{F}_t)$.

PROOF. a) It is immediate from the definition (4.3) that (τ_u) is increasing and right continuous. Additivity follows from the perfect additivity of L for $(\hat{\theta}_t)$.

b) The range of (τ_u) is M since M is a minimal right-closed set and L increases on M as in (4.2).

c) By the definition of (τ_u) and the left-continuity of L,

$$\{\tau_u < t\} = \{L_t > u\} \in \hat{F}_t$$

for all t and u. Since $(\hat{F}_t)$ is right continuous, this shows that each τ_u is a stopping time of $(\hat{F}_t)$.

Let T be a stopping time of (M_u) taking only countably many values u. Then,

$$\{\tau_T < t\} = \bigcup_u \{\tau_u < t, T = u\}.$$

Each term in the union belongs to $\hat{F}_t$, since τ_u is a stopping time of $(\hat{F}_t)$ and $\{T = u\} \in M_u \subset \hat{F}_{\tau_u}$. So, τ_T is a stopping time of $(\hat{F}_t)$.

The same conclusion holds with arbitrary stopping times T, because T can be approximated from above by T_n taking countably many values only, in which case each τ_{T_n} is a stopping time, (τ_{T_n}) decreases to τ_T by the right continuity of (τ_u), and $(\hat{F}_t)$ is right continuous.

(4.10) PROOF of Theorem (4.8). a) *Right continuity*. For (τ_u), this is in Lemma (4.9). Together with (4.9b) and the assumption of right continuity for X on M, this implies that Y is right continuous.

b) *Adaptedness*. Each τ_u is a stopping time of $(\hat{F}_t)$ and, therefore, is $\hat{F}_{\tau_u -}$ measurable. So, by (4.7), τ_u is M_u measurable. Of course, Y_u is so trivially by (4.7).

c) *Homogeneity*. Additivity of (τ_u) was mentioned in Lemma (4.9). Homogeneity of Y_u follows from it:

$$Y_v \circ \sigma_u = X_{\tau_v} \circ \hat{\theta}_{\tau_u} = X_{\tau_u + \tau_v \circ \hat{\theta}_{\tau(u)}} = X_{\tau_{u+v}} = Y_{u+v}$$

by the homogeneity condition (1.2iii) for X and by the fact that $[\tau_u] \subset M$ in view of (4.9b).

d) *Measurability* of $x \to \hat{P}_x(Y_u \in A, \tau_u \in B)$ for $A \in E$ and $B \in R_+$. The event $\{Y_u \in A, \tau_u \in B\}$ belongs to $\hat{G} = G \otimes F'$. Thus, it is sufficient to show that $x \to \hat{P}_x(\hat{G})$ is E-measurable for every $\hat{G} \in \hat{G}$. For $\hat{G}(\omega,\omega') = G(\omega)G'(\omega')$ the conclusion is immediate since $\hat{P}_x = P_x \times P'$

and since $x \to P_x(G)$ is $\mathcal{E}$-measurable by the way the P_x are chosen. The monotone class theorem concludes it for arbitrary $\hat{G}$ in $\hat{\mathcal{G}}$.

e) That (1.3) holds follows from (4.9b) and the way Y is defined. To complete the proof, we need to show the Markov and strong Markov properties for (Y,τ), namely, that

$$\hat{P}_x[Y_v \circ \sigma_U \in A,\ \tau_v \circ \sigma_U \in B \mid M_U] = \hat{P}_{Y_U}[Y_v \in A,\ \tau_v \in B]$$

for all $A \in \mathcal{E}$, $B \in \mathcal{R}_+$, and all stopping times U of (M_u). Since $\{Y_v \in A,\ \tau_v \in B\} \in \hat{\mathcal{G}}$ and $Y_v \circ \sigma_U = \Delta$ on $\{\tau_U = \infty\}$, this follows from the following proposition and completes the proof.

(4.11) PROPOSITION. *Let* $\hat{G} \in b\hat{\mathcal{G}}$ *and* U *a stopping time of* (M_u). *Then,*

$$\hat{E}_x[\hat{G} \circ \sigma_U \mid M_U] = \hat{E}_{Y_U}[\hat{G}] \quad \text{on} \quad \{\tau_U < \infty\}. \tag{4.12}$$

PROOF. Throughout the proof we set $T = \tau_U$. Then, $[T] \subset M$ and

$$Y_U = X_T, \qquad \{T < \infty\} = \{X_T \in E\}.$$

By (4.10a), Y is right continuous and adapted to (M_u). Thus, X_T is M_U-measurable. By Lemma (4.9c), T is a stopping time of $(\hat{F}_t)$.

a) On $\{X_T \in E\backslash C\}$, T is finite and belongs to $M\backslash M_i$, and $\sigma_U = \hat{\theta}_T$ by (4.5). So, recalling that $\hat{F}_T \supset M_U$ and using the regeneration property for (M,X) at the stopping time T (Theorem 3.10)), we get

$$\begin{aligned}\hat{E}_x[I_{\{X_T \in E\backslash C\}}\hat{G} \circ \sigma_U \mid M_U] &= I_{\{X_T \in E\backslash C\}}\hat{E}_x[\hat{E}_x[\hat{G} \circ \hat{\theta}_T \mid \hat{F}_T] \mid M_U] \\ &= I_{\{X_T \in E\backslash C\}}\hat{E}_x[\hat{E}_{X_T}(\hat{G}) \mid M_U] = I_{\{X_T \in E\backslash C\}}\hat{E}_{X_T}(\hat{G}).\end{aligned}$$

Thus, to show (4.12), there remains to show that

(4.13) $$\hat{E}_x[I_{\{X_T\in C\}}\,\hat{G}\circ\sigma_U \mid M_U] = I_{\{X_T\in C\}}\,\hat{E}_{X_T}(\hat{G}).$$

b) On $\{X_T \in C\}$, by Theorem (3.10), T is finite and belongs to M_i. But, $M_i = \bigcup_{j,k}[S_{jk}]$ by Lemma (3.6). Further,

$$\{X_T \in C,\ T = S_{jk}\} = (\{T > S_{jn};\ n < k\}\setminus\{T > S_{jn};\ n \le k\}) \cap \{\tfrac{1}{j} < \lambda(X_T) \le \tfrac{1}{j-1}\}$$

almost surely. Since the S_{jn} and T are stopping times of $(\hat{F}_t)$, $\{S_{jn} < T\} \in \hat{F}_{T-} \subset M_U$. So, to prove (4.13), it is enough to show that, for every j and k,

(4.14) $$\hat{E}_x[\hat{G}\circ\sigma_U \mid M_U] = \hat{E}_{X_T}(\hat{G}) \quad \text{on} \quad \{X_T \in C,\ T = S_{jk}\} \in M_U.$$

c) Fix j and k. By the monotone class theorem, it is sufficient to prove (4.14) for $\hat{G}$ having the form

(4.15) $$\hat{G} = G\cdot\dot{F}\cdot f\circ W_{j1}\cdot g\circ(W_{jn})_{n\ge 2}$$

where $G \in bG$, $f \in bR_+$, $g \in b(R_+)^{\mathbb{N}_0}$, and $\dot{F} \in b\dot{F}$ is such that it is free of the coordinates W_{jn}, $n \ge 1$. Then, by (4.5)-(4.6),

$$\hat{G}\circ\sigma_U = G\circ\hat{\theta}_T\cdot\dot{F}\circ\hat{\theta}_T\cdot f\circ\hat{W}_{jk}\cdot g\circ(W_{jn}\circ\hat{\theta}_T)_{n\ge 1} \quad \text{on} \quad \{T = S_{jk}\}$$

where

(4.16) $$\hat{W}_{jk} = W_{jk} - A_U/\lambda(Y_U) = W_{jk} - (U - L_T)/\lambda(X_T)$$

(recall that, on $\{T = S_{jk}\}$, L is left continuous at T and jumps at that isolated point T by the amount $W_{jk}/\lambda(X_T)$). Thus, by the regeneration property (Theorem 3.10) of (M,X) at T, since $f(\hat{W}_{jk})\, I_{\{T=S_{jk}\}}$ is $\hat{F}_T$ measurable,

$$\hat{E}_x[G\circ\sigma_U \mid \hat{F}_T] = f\circ\hat{W}_{jk}\ \hat{E}_x[(G\cdot \overset{\prime}{F}\cdot g\circ(W_{jn})_{n\geq 1})\circ\hat{\theta}_T \mid \hat{F}_T]$$

$$= f\circ\hat{W}_{jk}\ \hat{E}_{X_T}[G\cdot \overset{\prime}{F}\cdot g\circ(W_{jn})_{n\geq 1}] \quad \text{on} \quad \{T=S_{jk}<\infty\}.$$

Hence, since $\hat{F}_T \supset M_U$ and $X_T = Y_U$ is in M_U,

$$(4.17)\qquad \hat{E}_x[\hat{G}\circ\sigma_U \mid M_U] = \hat{E}_{X_T}[G\cdot \overset{\prime}{F}\cdot g\circ(W_{jn})_{n\geq 1}]\ \hat{E}_x[f\circ\hat{W}_{jk}\mid M_U]$$

on $\{T=S_{jk}<\infty\}$.

Suppose (for the time being) that

$$(4.18)\qquad \hat{E}_x[f\circ\hat{W}_{jk}\mid M_U] = \int_0^\infty f(x)e^{-x}\,dx = \overset{\prime}{E}[f\circ W_{j1}] \quad \text{on} \quad \{T=S_{jk}<\infty\}.$$

Putting this into (4.17) we obtain

$$(4.19)\qquad \hat{E}_x[\hat{G}\circ\sigma_U|M_U] = \hat{E}_{X_T}[G\cdot \overset{\prime}{F}\cdot g\circ(W_{jn})_{n\geq 1}]\ \overset{\prime}{E}[f\circ W_{j1}) \quad \text{on} \quad \{T=S_{jk}<\infty\}.$$

On the other hand, $\hat{P}_y = P_y\times\overset{\prime}{P}$, $\overset{\prime}{P}$ is a product measure, and $\overset{\prime}{F}$ is free of the W_{jn}, $n\geq 1$. So,

$$\hat{E}_y[G\cdot \overset{\prime}{F}\cdot g\circ(W_{jn})_{n\geq 1}]\ \overset{\prime}{E}[f\circ W_{j1}] = E_y(G)\,\overset{\prime}{E}(\overset{\prime}{F})\,\overset{\prime}{E}(g\circ(W_{jn})_{n\geq 1})\ \overset{\prime}{E}(f\circ W_{j1})$$

$$= E_y(G)\ \overset{\prime}{E}(\overset{\prime}{F})\ \overset{\prime}{E}(g\circ W_{jn})_{n\geq 2})\ \overset{\prime}{E}(f\circ W_{j1})$$

$$= E_y[G\cdot \overset{\prime}{F}\cdot f\circ W_{j1}\cdot g\circ(W_{jn})_{n\geq 2}] = \hat{E}_y(\hat{G}).$$

Putting this into (4.19) yields (4.14) and completes the proof assuming that (4.18) is true.

d) To complete the proof, there remains to show (4.18) with $\hat{W}_{jk}$ as in (4.16). Suppose U is a deterministic time, say $U = u$. Given M_u, on $\{T = S_{jk} < \infty\}$, the conditional distribution of $\hat{W}_{jk}$ is that of the difference $W_{jk} - A_u/\lambda(X_T)$ given that the exponentially distributed random variable W_{jk} exceeds the given value $A_u/\lambda(X_T)$. So, by the memorylessness of such variables, we have (4.18). This extends to U taking countably many values.

Next, suppose that the stopping time U is arbitrary. By the monotone class theorem, we may suppose (and do) that f is continuous. Approximate U from above by stopping times U_n of (M_u) taking values in $\{m/2^n;\ m \geq 1,\ n \geq 1\}$, and set

$$T_n = \tau(U_n), \quad W^n_{jk} = \hat{W}_{jk} - (U_n - U)/\lambda(X_T).$$

On $\{T = S_{jk} < \infty\}$, we have $L_T \leq U < L_{T+} = L_T + \lambda(X_T)\, W_{jk}$. Thus, as U_n decreases to U, we have T_n approaching T and $\hat{W}^n_{jk}$ approaching $\hat{W}_{jk}$. Thus, by the bounded convergence theorem, for $F \in bM_U$,

$$(4.20) \qquad \hat{E}_x[I_{\{T=T_n=S_{jk}<\infty\}}\, F \cdot f \circ W^n_{jk}] \to \hat{E}_x[I_{\{T=S_{jk}<\infty\}}\, F \cdot f \circ \hat{W}_{jk}].$$

But, by the arguments of the preceding paragraph, applied with U_n, the left side of (4.20) is equal to (note that $F \in bM_U \subset bM_{U_n}$),

$$\hat{E}_x[I_{\{T=T_n=S_{jk}<\infty\}}\, F] \cdot \dot{E}[f \circ W_{j1}],$$

which approaches $\hat{E}_x[I_{\{T=S_{jk}<\infty\}}\, F] \cdot \dot{E}[f \circ W_{j1}]$. Hence,

$$\hat{E}_x[I_{\{T=S_{jk}<\infty\}} \, f \circ \hat{W}_{jk} \mid M_U] = I_{\{T=S_{jk}<\infty\}} \, E'[f \circ W_{j1}].$$

This completes the proof of (4.18), and thus that of Proposition (4.11).

References

1. R.M. BLUMENTHAL and R.K. GETOOR. *Markov Processes and Potential Theory.* Academic Press, New York, 1968.

2. E. ÇINLAR. Markov additive processes, II. *Z. Wahrscheinlichkeitstheorie verw. Gebiete 31* (1972), 94-121.

3. E. ÇINLAR and J. JACOD. Representation of semimartingale Markov processes in terms of Wiener processes and Poisson random measures. *Seminar on Stochastic Processes 1981,* 159-242. Birkhäuser, Boston, 1981.

4. J. JACOD. Systèmes régénératifs et processus semi-markoviens. *Z. Wahrscheinlichkeitstheorie verw. Gebiete 31* (1974), 1-23.

5. H. KASPI. Excursions of Markov processes from a Borel set via Markov additive processes. To appear.

6. B. MAISONNEUVE. Systèmes régénératifs. *Astérisque, No. 15, Soc. Math. France,* Paris, 1974.

7. B. MAISONNEUVE. Ensembles régénératifs, temps locaux et subordinateurs. *Séminaire de Probabilités V (Univ. Strasbourg),* 147-169. *Lecture Notes Math. 191.* Springer-Verlag, Berlin, 1971.

E. ÇINLAR
IE/MS Department
Northwestern University
Evanston, IL 60201 USA

H. KASPI
IE Department
Technion
Haifa, Israel

Seminar on Stochastic Processes, 1982
Birkhäuser, Boston, 1983

EXCURSIONS AND FORWARD TIMES*

by

R. K. GETOOR

1. Introduction

This paper continues the work of [3] and [4]. As in those papers, for the principal results, one assumes given a pair of standard processes X and $\hat{X}$ with dual density relative to some σ-finite measure on their common state space E. The main results of [3] and [4] concerned the construction and properties of a family of measures $P^{x,\ell,y}$ which were shown to "govern" the distribution of the excursions of X from a given closed homogeneous optional set M, conditional on the excursion starting at x, ending at y, and having length ℓ. The meaning of "govern" in the above statement was made precise for excursions straddling an arbitrary stopping T in [3] and [4].

In the present paper, we study this question for excursions straddling a class of times τ that we call *forward* times. Roughly speaking, a forward time τ is one for which the event $\{\tau > t\}$ depends on the process X_s only for times $s \geq t$ for each $t \geq 0$. See §2 for the precise definition. In [1] and [5] such times were called "backward" times, but we feel the name *forward* time is more

* This research was supported in part by NSF Grant MCS79-23922

appropriate. In an obvious sense, forward times are dual to stopping times. Clearly, co-optional times are forward times. In fact, in [2] it was shown that any forward time for X is (appropriately interpreted) a co-optional time for a space-time process over X.

In §2 we develop some elementary properties of forward times that will be needed in the later sections. The main results on excursions straddling a forward time τ are given in §3 and §4. There are no surprises except, perhaps, that it is so simple. The basic facts are contained in (3.7) and (4.9). There is a certain simplification if τ is co-optional, and this is made explicit in (3.10) and the remarks following (4.9). By using the "extended" exit system of [7] we are able to allow $G = 0$ in our theorems. This extension also applies to the results in [3] and [4], and we shall use this extension without comment.

In the special case of an exact coterminal time τ one may obtain stronger results. These are spelled out in §5. Perhaps the most interesting observation is the following consequence of Theorem 5.11 and Theorem 13.7 of [3]. If T is the exact terminal time corresponding to τ, then the law of the excursion straddling T conditional on what happens outside the excursion interval is exactly the same as the law of the excursion straddling τ, conditional on the analogous quantities. This is made precise following (5.11). For example, if τ is the last exit time from a Borel set B, then T is the first hitting time of B. The fact that the appropriately conditioned excursions straddling T and τ have the same distributions may be made intuitively plausible by a time reversal argument using (vii) of Theorem 7.6 in [3].

2. Forward Times

Let E be a Lusinian state space and let Ω consist of all right continuous paths $\omega : \mathbb{R}^+ \to E_\delta \equiv E_\delta \cup \{\delta\}$ where δ is a cemetery point not in E. That is, δ is adjoined to E as an isolated point and if $\zeta(\omega) \equiv \inf\{t : \omega(t) = \delta\}$ then $\omega(t) = \delta$ for all $t \geq \zeta(\omega)$. Here and in the sequel "$\equiv$" is the symbol for "is defined to be." As usual $X_t(\omega) \equiv \omega(t)$. Beginning in §4 we shall assume, in addition, that each $\omega \in \Omega$ has left limits in E on $]0,\zeta(\omega)[$. We shall use the standard notation $\mathcal{F}^0 = \sigma\{X_t : t \geq 0\}$ for the σ-algebra on Ω generated by the coordinate maps X_t when E is furnished with its Borel σ-algebra $\mathcal{E}$. Also $\mathcal{F}^0_t = \sigma\{X_s : s \leq t\}$. If $(A,\mathcal{A})$ is any measurable space, $\mathcal{A}^*$ denotes the σ-algebra of universally measurable sets over $(A,\mathcal{A})$. However, in keeping with the usual abuse of notation, $\mathcal{F}^* = (\mathcal{F}^0)^*$ and $\mathcal{F}^*_t = \bigcap_Q (\mathcal{F}_t \vee \mathcal{N}^Q)$ where, for Q a finite measure on $(\Omega,\mathcal{F}^0)$, $\mathcal{N}^Q$ is the ideal of all Q null sets (in the completion of $\mathcal{F}^0$ under Q), and the intersection is over all finite measures Q on $(\Omega,\mathcal{F}^0)$. In this section we shall deal only with algebraic properties and so we delay until §3 the introduction of measures under which X is Markovian.

The shift operators $\theta_t : \Omega \to \Omega$ and the killing operators $k_t : \Omega \to \Omega$ are defined as usual for $t \geq 0$ by $(\theta_t\omega)(s) = \omega(t + s)$ and $k_t\omega(s) = \omega(s)$ if $s < t$, $= \delta$ if $s \geq t$. Finally we need the splicing map $(\omega',t,\omega) \to \omega'/t/\omega$ given by $X_s(\omega'/t/\omega) = X_s(\omega')$ if $s < t$ and $X_s(\omega'/t/\omega) = X_{s-t}(\omega)$ if $s \geq t$. We record some elementary identities relating these operations:

(2.1) (i) $\theta_t k_{t+s} = k_s \theta_t$;

(ii) $\theta_t(\omega'/s/\omega) = \theta_t\omega'/s-t/\omega$ if $t < s$,

$= \theta_{t-s}\omega$ if $t \geq s$;

(iii) $k_t(\omega'/s/\omega) = k_t\omega'$ if $t \le s$,

$= \omega'/s/k_{t-s}\omega$ if $t > s$;

(iv) $\omega'/s/\omega = k_s\omega'/s/\omega$.

It is immediate that $(\omega',t,\omega) \to \omega'/t/\omega$ is $(F^0 \times B^+ \times F^0)/F^0$ measurable where B^+ is the σ-algebra of Borel sets of $\mathbb{R}^+ = [0,\infty[$. Consequently the splicing map is $(F^0 \times B^+ \times F^0)^*/F^*$ measurable.

For each $t \ge 0$ define

(2.2) $$G_t^0 = \theta_t^{-1} F^0 \quad ; \; G_t^* = \theta_t^{-1} F^*.$$

These σ-algebras decrease as t increases and represent the "future" at t just as F_t^0 and F_t^* represent the past.

(2.3) DEFINITION. *A forward time* τ *is a map from* Ω *to* $[0,\infty]$ *such that* $\{\tau > t\} \in G_t^*$ *for each* $t \ge 0$.

Clearly τ is F^* measurable.

(2.4) REMARK. In the Markovian situation it is easy to see that if $\tau : \Omega \to [0,\infty]$ satisfies $\{\tau > t\} \in \theta_t^{-1}F$ for each $t \ge 0$ where F has its usual meaning, then there exists a forward time τ^* as defined in (2.3) with $\tau^* = \tau$ almost surely.

A *co-optional* time is a map $\tau : \Omega \to [0,\infty]$ which is F^* measurable and satisfies for each $t \ge 0$

(2.5) $$\tau\circ\theta_t = (\tau - t)^+$$

identically in ω. A co-optional time τ is a forward time because $\{\tau > t\} = \{\tau\circ\theta_t > 0\}$. We shall also consider coterminal times in §5, but we postpone a formal definition until needed.

(2.6) LEMMA. *Let* τ *be forward time. Then,*

(i) $\tau\, 1_{\{\tau > s\}} \in G_s^*$;

(ii) *if* $\tau(\omega) > s$ *and* $\theta_s\omega = \theta_s\omega'$, *then* $\tau(\omega) = \tau(\omega')$.

PROOF. The first assertion is immediate from

$$\tau\, 1_{\{\tau > s\}} = \lim_n \sum_{k=0}^{\infty} s(n,k) 1_{]s(n,k),\, s(n,k+1)]}(\tau) + \infty\, 1_{\{\tau = \infty\}} ,$$

where $s(n,k) = s + k2^{-n}$. But (ii) is evident from (i).

We now introduce some transformations of a forward time that will be of central importance in the sequel. If τ is a forward time define for $s \geq 0$

$$\tau_s(\omega) = (\tau(\omega_0/s/\omega)-s)^+ \tag{2.7}$$

where ω_0 is a fixed but arbitrary point in Ω. Since $\theta_s(\omega_0/s/\theta_s\omega) = \theta_s\omega$, it follows from (2.6) that

$$\tau_s(\theta_s\omega) = (\tau(\omega)-s)^+. \tag{2.8}$$

More generally, for $s \geq 0$ and $r \geq 0$,

$$\tau_{s+r}(\theta_s\omega) = (\tau_r(\omega)-s)^+ , \tag{2.9}$$

because

$$\theta_{s+r}(\omega_0/r/\omega) = \theta_s\omega = \theta_{s+r}(\omega_0/s+r/\theta_s\omega).$$

REMARK. If we define $\tilde{\tau}(s,\omega) = \tau_s(\omega)$ on $\mathbb{R}^+ \times \Omega$ and $\tilde{\theta}_t(s,\omega) = (s+t, \theta_t\omega)$ from $\mathbb{R}^+ \times \Omega \to \mathbb{R}^+ \times \Omega$, then (2.9) states that $(\tilde{\tau}-s)^+ = \tilde{\tau}\circ\tilde{\theta}_s$. That is, $\tilde{\tau}$ is co-optional with respect to the shifts $(\tilde{\theta}_t)$. See (6.3) of [2].

If τ is co-optional, then since $\theta_s(\omega_0/s/\omega) = \omega$, $\tau_s(\omega) = \tau(\omega)$ for all $s \geq 0$ and (2.8) reduces to (2.5).

In §4 we shall need one more quantity. Define J on $\mathbb{R}^+ \times \Omega \times \mathbb{R}^+ \times \Omega$ by

$$(2.10) \qquad J(t,\omega,s,\omega') \equiv [\ \tau((\omega_0/s/\omega')/t+s/\omega)-s\ ]^+ = \tau_s(\omega'/t/\omega).$$

The last equality obtains because $J(t,\omega,s,\omega') = \tau_s(\theta_s w)$ where $\mathbf{w} = (\omega_0/s/\omega')/t+s/\omega$ and $\theta_s w = \omega'/t/\omega$. Next observe that

$$(2.11) \qquad J(t,\theta_t\omega,s,k_t\omega) = \tau_s(k_t\omega/t/\theta_t\omega) = \tau_s(\omega).$$

This is a representation of $\tau_s(\omega)$ in terms of the part of the path before $t, k_t\omega$ and the part after $t, \theta_t\omega$.

3. Excursions Straddling Forward Times

Let $(P^x, x \in E)$ be a family of probabilities on $(\Omega, \mathcal{F}^*)$ which makes the coordinate maps (X_t) a right process. If the reader prefers, he may suppose that the right process $X = (X_t, P^x)$ is, in fact, a standard process. We shall specialize to standard processes in duality beginning in §4.

We now fix an optional set $M \subset]0,\zeta[$ which is homogeneous on $]0,\infty[$ and closed in $]0,\zeta[$. See [6], [7], or [3]. Meyer has shown that we may assume that

$$(3.1) \qquad R \equiv \inf\{t > 0 : t \in M\}$$

is F^* measurable. Of course, R is a (perfect, exact) terminal time. Let M_ℓ be the set of strictly positive left endpoints of the intervals contiguous to M. Define

$$M_\ell^0(\omega) = M_\ell(\omega) \text{ if } R(\omega) = 0, \quad M_\ell^0(\omega) = M_\ell(\omega) \cup \{0\} \text{ if } R(\omega) > 0.$$

Then there exists an extended exit system $({}^*P^x, B^0)$. See page 64 of [7]. Here $dB_t^0 \equiv dB_t + \varepsilon_0(dt)\, 1_{\{R > 0\}}$ where B is an additive functional of X with a bounded 1-potential and ${}^*P^x(d\omega)$ is a kernel from (E,E^*) to (Ω,F^*). If $Z \geq 0$ is an optional process and $F \geq 0$ is $B^+ \times E \times F^0$ measurable, then

$$(3.2) \quad E^x \sum_{s \in M_\ell^0} Z_s\, F(s,X_s,\theta_s) = E^x \int_{[0,\infty[} Z_s \int F(s,X_s,\omega')\, {}^*P^{X(s)}(d\omega')dB_s^0.$$

Moreover, for each $x \in E$, ${}^*P^x$ is σ-finite, ${}^*P^x(\zeta = 0) = 0$, ${}^*P^x(R = 0) = 0$, $0 < {}^*P^x(1-e^{-R}) \leq 1$, and if $P^x(R > 0) = 1$ then ${}^*P^x = P^x$. The properties of exit systems are discussed in [6] and [4], and are easily extended to the "extended" exit system considered here.

REMARK. A standard completion argument shows that (3.2) is valid for $F \geq 0$ in $(B^+ \times E \times F^0)^*$. In particular, for such an F, $(s,x) \to I(s,x) \equiv \int F(s,x,\omega')\, {}^*P^x(d\omega')$ is in $(B^+ \times E)^*$ and

$(s,\omega) \to I(s,X_s(\omega))$ is in $(\mathcal{B}^+ \times \mathcal{F}^0)^*$. As a result $\int I(s,X_s(\omega))dB_s^0(\omega)$ exists for each ω and is $\mathcal{F}^*$ measurable. In other words, all the integrals on the right side of (3.2) make sense if $F \geq 0$ is $(\mathcal{B}^+ \times \mathcal{E} \times \mathcal{F}^0)^*$ measurable. In the sequel we shall omit such routine measurability assertions.

We now fix a forward time τ. Define

(3.3) $$G = G_\tau = \sup\{t \leq \tau : t \in M\}$$
$$D = D_\tau = \inf\{t > \tau : t \in M\} = \tau + R\circ\theta_\tau .$$

The interval $]G,D[$ is called the excursion interval straddling τ. Note that $D = G + R\circ\theta_G$, and if $G < \tau < \infty$, then $\tau < D$ since M is closed. Also observe that ($D_s = s + R\circ\theta_s$ for all $s \in \mathbb{R}^+$)

$$s = G < \tau < \infty \iff s \in M_\ell^0 , \quad s < \tau < D_s$$
$$\iff s \in M_\ell^0 , \quad 0 < \tau_s\circ\theta_s < R\circ\theta_s$$

where τ_s is defined in (2.7) and satisfies $(\tau-s)^+ = \tau_s\circ\theta_s$. The following theorem is now an immediate consequence of (3.2) and the remark following. $\mathcal{O}$ denotes the optional σ-algebra.

(3.4) THEOREM. *Let* $Z \in b\mathcal{O}$, $F \in bF^*$, $f \in b(\mathcal{E} \times \mathcal{B}^+)^*$. *Then for each initial probability* μ

(3.5) $$E^\mu[\, Z_G\, f(X_G,G)\, F\circ\theta_G;\, G < \tau < \infty\,]$$
$$= E^\mu \int Z_s\, f(X_s,s)\, {}^*P^{X(s)}[F;\, 0 < \tau_s < R]\, dB_s^0 .$$

REMARKS. Clearly the left side of (3.5) is finite, and hence, so is the right side. It follows from the definition (2.7) of τ_s that $(s,\omega) \to \tau_s(\omega)$ is $(\mathcal{B}^+ \times \mathcal{F}^0)^*$ measurable. Hence, $(s,\omega) \to F(\omega)\, 1_{\{0 < \tau_s(\omega) < R(\omega)\}}$ is in $(\mathcal{B}^+ \times \mathcal{F}^0)^*$, and $(s,x) \to q(s,x) \equiv {}^*P^x[F;\ 0 < \tau_s < R]$ is in $(\mathcal{B}^+ \times \mathcal{E})^*$. Therefore $(s,\omega) \to q(s,X_s(\omega))$ is in $(\mathcal{B}^+ \times \mathcal{F}^0)^*$.

For $x \in E$ and $s \geq 0$ define a measure $H^{x,s}$ on $\mathcal{F}^*$ by

$$(3.6) \qquad H^{x,s}(F) = \frac{{}^*P^x[\ F\ ;\ 0 < \tau_s < R\ ]}{{}^*P^x[\ 0 < \tau_s < R\ ]}$$

where $0/0 = 0$ and $\infty/\infty = 0$. Clearly $H^{x,s}$ is either a probability or the zero measure according as $h(x,s) \equiv {}^*P^x(0 < \tau_s < R)$ satisfies $0 < h(x,s) < \infty$ or not.

If $S \geq 0$ is $\mathcal{F}$ measurable, then $\mathcal{F}_S$ is the σ-algebra defined by $Z \in \mathcal{F}$ and for each initial measure μ there exists an optional process (Z_t^μ) with $Z = Z_S^\mu$ almost surely P^μ on $\{S < \infty\}$. This differs slightly from the definition of $\mathcal{F}_S$ given in [3]. The above defined $\mathcal{F}_S$ is the intersection over all μ of the P^μ completion of the one defined in [3].

Since M is optional, G and X_G are $\mathcal{F}_G$ measurable. If $F \in b\,\mathcal{F}^*$, then $(x,s) \to H^{x,s}(F)$ is in $(\mathcal{E} \times \mathcal{B}^+)^*$, and it follows readily from these observations that as a function of ω, $H^{X(G),G}(F)$ is in $\mathcal{F}_G$.

(3.7) COROLLARY. *Let* $Z \in b\mathcal{O}$, $F \in b\,\mathcal{F}^*$. *Then*

$$(3.8) \qquad E^\mu[\ Z_G\ F\circ\theta_G;\ G < \tau < \infty\ ] = E^\mu[\ Z_G\ H^{X(G),G}(F)\ ;\ G < \tau < \infty\ ].$$

If $h(x,s) = {}^*P^x[0 < \tau_s < R]$, *then* $0 < h(X_G,G) < \infty$ *almost surely on* $\{G < \tau < \infty\}$.

REMARK. It is immediate from (3.8) that

$$(3.9) \qquad E^{\mu}[\, F\circ\theta_G \mid \mathcal{F}_G \,] = H^{X(G),G}(F) \quad \text{on} \quad \{G < \tau < \infty\}.$$

Thus $H^{x,s}$ is the law of the post G process conditional on $\mathcal{F}_G, X(G) = x$, and $G = s$ on $\{G < \tau < \infty\}$.

PROOF. Let $Z \in b\mathcal{O}$ and $F \in b\mathcal{F}^*$. Using (3.5) twice and the fact h defined below (3.8) is in $b(\mathcal{E} \times \mathcal{B}^+)^*$ gives

$$\begin{aligned} E^{\mu}[\, Z_G\, F\circ\theta_G;\, G < \tau < \infty \,] &= E^{\mu} \int Z_s\, h(X_s,s)\, H^{X(s),s}(F)\, dB^0_s \\ &= E^{\mu}[\, Z_G\, H^{X(G),G}(F);\, G < \tau < \infty \,]. \end{aligned}$$

For the second assertion let $N_0 = \{h = 0\} \in (\mathcal{E} \times \mathcal{B}^+)^*$. Then by (3.5)

$$E^{\mu}[1_{N_0}(X_G,G); G < \tau < \infty] = E^{\mu} \int 1_{N_0}(X_s,s)\, h(X_s,s)\, dB^0_s = 0.$$

If $N_\infty = \{h = \infty\}$, then

$$1 \geq E^{\mu}[1_{N_\infty}(X_G,G); G < \tau < \infty] = E^{\mu} \int 1_{N_\infty}(X_s,s)\, h(X_s,s)\, dB^0_s.$$

But the value of the last integral is either zero or infinity and consequently must vanish. This completes the proof of (3.7).

In case τ is a co-optional time τ_s and $H^{x,s}$ are independent of s. We record this important case explicitly.

(3.10) COROLLARY. *Let* τ *be a co-optional time and define* $H^x(F) = {}^*P^x[F \mid 0 < \tau < R]$. *Then*

$$E^{\mu}[\,F\circ\theta_G \mid F_G\,] = H^{X(G)}(F)$$

on $\{G < \tau < \infty\}$ *and* $0 < {}^{*}P^{X(G)}[0 < \tau < R] < \infty$ *almost surely on* $\{G < \tau < \infty\}$.

4. Conditional Excursions: Forward Times

In the remainder of this paper we suppose that X and $\hat{X}$ are standard processes on a Lusinian state space E having dual densities relative to a σ-finite measure $\xi(dx) = dx$ on E. We take Ω to be the set of paths $\omega : \mathbb{R}^+ \to E \cup \{\delta\}$ which are right continuous on $\mathbb{R}^+$ and have left limits in E on $]0,\zeta(\omega)[$. These assumptions are discussed in [3]. As in §3, M is a closed homogeneous optional set contained in $]0,\zeta[$. Under the present hypotheses one may suppose that M is $B^+ \times F^0$ measurable (see §4 of [3]). Moreover the kernel ${}^{*}P^{x}$ appearing in the extended exit system may be chosen to be a kernel from (E,E) to (Ω,F^0). In [3] a family of measures $P^{x,\ell,y}$ on Ω for $x, y \in E$, $\ell > 0$ was constructed and it was shown in [4] how these measures govern the excursion straddling a stopping time. In this section we shall consider excursions straddling a forward time. It was shown in [3] that each $P^{x,\ell,y}$ is either a probability or zero, and $(x,\ell,y) \to P^{x,\ell,y}(F)$ is $E \times B^+ \times E$ measurable for $F \in b\,F^0$.

Fix a forward time τ and define $G = G_\tau$, $D = D_\tau$ as in §3. Because M is in $B^+ \times F^0$ and $\tau \in F^*$, G and D are in F^* under our present hypotheses. Recall the definition $J(t,\omega,s,\omega')$ in (2.10) and the representation (2.11) for τ_s in terms of J. The next lemma is the key to our discussion.

(4.1) LEMMA. *Let* $F \in b\,F^*$ *and* $\Psi \in b(B^+ \times E \times F^0)^*$. *Then for each*

$x \in E$ *and* $s \geq 0$

$$(4.2)\qquad {}^{*}P^{x}[\ F\circ k_R\ \Psi(R, X_{R-}, \theta_R);\ 0 < \tau_s < R,\ R < \infty\ ]$$

$$= {}^{*}P^{x}[\ K_s^{x,R,\ X_{R-},\theta_R}(F)\ \Psi(R, X_{R-}, \theta_R);\ R < \infty\ ]$$

where

$$(4.3)\qquad K_s^{x,r,y,\omega}(F) = \int F(\omega')1_{\{0 < J(r,\omega,s,\omega') < r\}}P^{x,r,y}(d\omega').$$

PROOF. It is a routine matter to check, using (2.10) and the fact that the splicing map is $(\mathcal{F}^0 \times \mathcal{B}^+ \times \mathcal{F}^0)^*/\mathcal{F}^*$ measurable, that J is in $(\mathcal{B}^+ \times \mathcal{F}^0 \times \mathcal{B}^+ \times \mathcal{F}^0)^*$. Let

$$(4.4)\qquad G(t,\omega,s,\omega') = 1_{\{0 < J(t,\omega,s,\omega') < t\}}.$$

Then, by (2.11), for each $t \geq 0$,

$$\{\ 0 < \tau_s(\omega) < t\ \} = \{\ G(t, \theta_t\omega, s, k_t\omega) = 1\ \},$$

and so

$$(4.5)\qquad \{\ 0 < \tau_s < R\ \} = \{\ G(R, \theta_R, s, k_R) = 1\ \}.$$

In proving (4.2) it suffices to consider Ψ of the form $\Psi(r,x,\omega) = \varphi(r,x)Y(\omega)$ with $\varphi \in b(\mathcal{B}^+ \times \mathcal{E})$ and $Y \in b\,\mathcal{F}^0$. For fixed x define finite measures μ and ν on $\mathcal{B}^+ \times \mathcal{F}^0 \times \mathcal{F}^0$ by

$$(4.6)\qquad \mu(H) = {}^{*}P^{x}[\ H(R, \theta_R, k_R)(1-e^{-R});\ R < \infty\ ]$$

$$\nu(H) = {}^{*}P^{x}[\ (1-e^{-R}) \int H(R, \theta_R, \omega')P^{x,R,X_{R-}}(d\omega')\ ].$$

For s fixed, choose $H \in b(\mathcal{B}^+ \times \mathcal{F}^0 \times \mathcal{F}^0)$ such that $H(t,\omega,\omega') =$

$G(t,\omega,s,\omega')$ almost everywhere with respect to both μ and ν. Since ${}^*P^x(R = 0) = 0$, in view of (4.5), the left hand side of (4.2) with Ψ as in the first sentence of this paragraph is equal to

$$(4.7) \qquad I = {}^*P^x[\ F\circ k_R\ \varphi(R,X_{R-})\ Y\circ\theta_R\ H(R,\theta_R,k_R);\ R < \infty\ ].$$

If $H(r,\omega,\omega') = h(r)a(\omega)b(\omega')$ with $h \in b\,\mathcal{B}^+$ and $a,\ b \in b\,\mathcal{F}^0$, then using the strong Markov property of ${}^*P^x$,

$$I = {}^*P^x[\ (Fb)\circ k_R\ \varphi(R,X_{R-})\ h(R)\ E^{X_R}(Ya);\ R < \infty\ ].$$

Next we use (3.10) of [4] and the strong Markov property again to conclude

$$I = {}^*P^x[\ P^{x,R,X_{R-}}(Fb)\ \varphi(R,X_{R-})\ h(R)\ (Ya)\circ\theta_R;\ R < \infty\ ].$$

Hence by the monotone class theorem it follows that I as originally written in (4.7) becomes

$${}^*P^x[\ \int F(\omega')\ H(R,\theta_R,\omega')\ P^{x,R,X_{R-}}(d\omega')\ \varphi(R,X_{R-})\ Y\circ\theta_R;\ R < \infty\ ].$$

Recalling the definition of ν in (4.5) and the choice of H one may replace $H(R,\theta_R,\omega')$ by $G(R,\theta_R,s,\omega')$ in this last expression and taking into account (4.3) and (4.4) this establishes (4.2).

We now come to the main result of this section. Let $L = D - G$ for $G < \infty$ be the length of the excurstion straddling τ. Clearly $L = R\circ\theta_G$ if $G < \infty$. Let

$$(4.8) \qquad \Lambda = \{\ G < \tau < \infty,\ D < \infty\ \}$$

and note that $\Lambda = \{G < \tau < \infty;\ R\circ\theta_G < \infty\}$. Also recall that $F_{\geq D}$ is the σ-algebra defined by $H \in F_{\geq D}$ provided $H \in F$ and for each μ there exists $\overline{H} \in F^*$ with $H = \overline{H}\circ\theta_D$ a.s. P^μ on $\{D < \infty\}$.

(4.9) THEOREM. *Let* $F \in b\,F^*$. *Then on* Λ

$$(4.10) \qquad E^\mu[\,F\circ k_R\circ\theta_G \mid F_G, F_{\geq D}, L, X_{D-}\,] = Q^{X_G,G,L,X_{D-},\theta_D}(F),$$

where

$$(4.11) \qquad Q^{x,s,\ell,y,\omega}(F) = \frac{K_s^{x,\ell,y,\omega}(F)}{K_s^{x,\ell,y,\omega}(1)}$$

Here $0/0 = 0$ *and* K *is defined in* (4.3). *Moreover* $K_G^{X_G,L,X_{D-},\theta_D}(1) > 0$ *almost surely on* Λ, *and if*

$$N = \{\ (x,s,\ell,y,\omega): Q^{x,s,\ell,y,\omega}(X_0 = x,\ \zeta = \ell,\ X_{\zeta-} = y) = 1\ \},$$

then $(X_G,G,L,X_{D-},\theta_D) \in N$ *almost surely on* Λ.

REMARKS. From the defintion of K in (4.3) and J in (2.10) we see that

$$(4.12) \qquad Q^{x,s,\ell,y,\omega}(F) = P^{x,\ell,y}[\ F \mid 0 < \tau_s(\cdot\,/\ell/\omega) < \ell\ ].$$

In particular if τ is a co-optional time τ_s is independent of s. Hence defining

$$(4.13) \qquad Q^{x,\ell,y,\omega}(F) = P^{x,\ell,y}[\ F \mid 0 < \tau\,(\cdot/\ell/\omega) < \ell\ ],$$

the right side of (4.10) reduces to $Q^{X_G,L,X_{D-},\theta_D}(F)$ and the explicit dependence on G disappears. This should be compared with Theorem 3.5 in [4]. It follows from (4.3) and the first sentence of the proof of (4.1) that $Q^{x,s,\ell,y,\omega}(F)$ is $(\mathcal{E} \times \mathcal{B}^+ \times \mathcal{B}^+ \times \mathcal{E} \times \mathcal{F}^0)^*$ measurable. Under the present hypotheses G, L, D, X_G, X_{D-} are all $\mathcal{F}^*$ measurable and so the right side of (4.10) is in $\sigma(\mathcal{F}_G,\mathcal{F}_{\geq D},L,X_{D-})^*$. This is the measurability content of (4.10).

PROOF. Given Lemma 4.1, the proof of Theorem 4.9 is similar to arguments used in [3] and [4], and so we omit some of the details in what follows. Let $Z \in b\,\mathcal{F}_G$, $\varphi \in b(\mathcal{B}^+ \times \mathcal{E})^*$, and $Y, F \in b\,\mathcal{F}^*$. Then using (3.8), $D = G + R\circ\theta_G$, and $L = R\circ\theta_G$

$$(4.14)\qquad I \equiv E^\mu\{\, Z\, F\circ k_R\circ\theta_G\, \varphi(L,X_{D-})\, Y\circ\theta_D;\ \Lambda\,\}$$

$$= E^\mu\{\, Z\, H^{X(G),G}[\, F\circ k_R\, \varphi(R,X_{R-})\, Y\circ\theta_R;\ R<\infty\,]:\ G<\tau<\infty\}.$$

From (4.2) and the definition (3.6) of $H^{x,s}$ we find

$$H^{x,s}[\, F\circ k_R\, \varphi(R,X_{R-})\, Y\circ\theta_R;\ R<\infty\,]$$

$$= H^{x,s}[Q^{x,s,R,X_{R-},\theta_R}\, \varphi(R,X_{R-})\, Y\circ\theta_R;\ R<\infty].$$

Substituting this into (4.14) and using (3.8) once again gives

$$I = E^\mu\{\, Z\, Q^{X_G,G,L,X_{D-},\theta_D}(F)\, \varphi(L,X_{D-})\, Y\circ\theta_D;\ \Lambda\,\}.$$

In view of the measurability discussion in the remarks following the statement of (4.9) this establishes (4.10).

The remaining two assertions in Theorem 4.9 are proved by the same arguments as those used in the proofs of the corresponding statements in Theorem 3.5 of [4]. We shall not repeat the details here.

5. Conditional excursions: Coterminal Times

A (perfect) *coterminal* time τ is a co-optional time that satisfies, in addition,

(5.1) (i) $\tau \circ k_t = \tau$ on $\{\tau < t\}$

(ii) $\tau \circ k_t \le t$

for all $t \ge 0$. This is the original definition of [8] except that we require τ to be F^* measurable. It follows that $\tau \circ k_t$ is F_t^* measurable and that

(5.2) $\tau \circ k_t \le \tau$ and $t \to \tau \circ k_t$ is increasing.

See [8] for (5.2). One says that τ is exact if

$$\tau = \sup_t \tau \circ k_t = \lim_{t \to \infty} \tau \circ k_t .$$

It was shown in [1] that if τ is exact then there exists a closed optional homogeneous set H such that $\tau = \sup H$. Then $T = \inf\{t > 0: t \in H\}$ is an exact (perfect) terminal time and

(5.3) $$T = \inf\{t > 0 : \tau \circ k_t > 0\}.$$

By Proposition 4.2 of [8], if τ is exact, then

$$(5.4) \qquad a < \tau \le b \iff T\circ\theta_a \le b - a,\ T\circ\theta_b = \infty ,$$

for $0 \le a < b < \infty$.

We now fix an exact coterminal time τ and let T be the corresponding terminal time. We assume without loss of generality the existence of a closed homogeneous optional set $H \subset]0,\zeta[$ with $\tau = \sup H$ and $T = \inf H$. Then $\{\tau > 0\} = \{T < \zeta\}$. Because $t + T\circ\theta_t$ is increasing and right continuous,

$$(5.5) \qquad Z_t = \lim_{s\uparrow\uparrow t} T\circ\theta_t , \quad t > 0,$$

exists and is left continuous on $]0,\infty[$. Clearly $Z_t\circ\theta_s = Z_{t+s}$ if $t > 0$, $s \ge 0$. Hence $Z \in \mathcal{H}^g$. Recall (see [2] or [9]) that $\mathcal{H}^g$ is the σ-algebra generated by left continuous processes which are perfectly homogeneous on $]0,\infty[$. Obviously $(t,\omega) \to Z_t(\omega)$ is in $\mathcal{B}^+ \times \mathcal{F}^*$ since T is $\mathcal{F}^*$ measurable in the present situation.

(5.6) LEMMA. $\{0 < \tau < t\} = \{T < t,\ Z_t = \infty\}$.

PROOF. Using (5.4)

$$\{0 < \tau < t\} = \bigcup_n \{0 < \tau \le t-1/n\}$$

$$= \bigcup_n \{T \le t-1/n,\ T\circ\theta_{t-1/n} = \infty\} \subset \{T < t,\ Z_t = \infty\}.$$

If $Z_t = \infty$, then $T\circ\theta_t = \infty$ because $t + T\circ\theta_t \ge t + Z_t$. Therefore $\{T < t,\ Z_t = \infty\} \subset \{0 < \tau \le t\}$. But if $t = \tau = \sup H$, then for $s < t$

$$s + T\circ\theta_s = \inf\{ u > s : u \in H \} \leq t.$$

Letting s increase to t, this implies that $Z_t = 0$, establishing (5.6).

Since a coterminal time τ is a co-optional time the results of §4, in particular (4.9) and (4.13), apply to τ. However, we shall obtain sharper results for coterminal times. The assumptions and notation are as in §4 except that τ is a coterminal time.

We shall need the *left* Markov property at the exact terminal time R. The precise statement is as follows: Given $Z \in b\,\mathcal{H}^g$ there exists $f \in b\,\mathcal{E}$ such that for all $\varphi \in b(\mathcal{B}^+ \times \mathcal{E})$ and $F \in b\,\mathcal{F}^0$ one has

$$(5.7)\quad E^\mu[\, Z_R\, \varphi(R,X_{R-})\, F\circ k_R;\ 0<R<\zeta\,] = E^\mu[\, f(X_{R-})\, \varphi(R,X_{R-})\, F\circ k_R;\ 0<R<\zeta\,].$$

See (5.2), (5.3), and the last paragraph of §2 of [2]. In the present situation $R \in \mathcal{F}^*$ and so (5.7) remains valid for $F \in b\,\mathcal{F}^*$, $\varphi \in b(\mathcal{B}^+ \times \mathcal{E})^*$. The next lemma extends (5.7) to the measures ${}^*P^x$.

(5.8) LEMMA. *Given* $Z \in b\,\mathcal{H}^g$, *let* f *correspond to* Z *as in* (5.7). *Then*

$${}^*P^x[\, Z_R\, F\circ k_R\, \varphi(R,X_{R-});\ R < \zeta\,] = {}^*P^x[\, f(X_{R-})\, \varphi(R,X_{R-})\, F\circ k_R;\ R < \zeta\,]$$

for $\varphi \in b(\mathcal{B}^+ \times \mathcal{E})^*$ *and* $F \in b\,\mathcal{F}^*$.

PROOF. Fix $t > 0$. Let $F \in b\,\mathcal{F}^0$. Then $F = h(g\circ\theta_t)$ with $h \in b\,\mathcal{F}^0_t$ and $g \in b\,\mathcal{F}^0$. If $t < R < \infty$, then $h\circ k_R = h$, $g\circ\theta_t\circ k_R = g\circ\theta_t\circ k_{t+R\circ\theta_t} = g\circ k_R\circ\theta_t$, and $Z_R = Z_{t+R\circ\theta_t} = Z_R\circ\theta_t$. Hence

$$*P^x\{ Z_R\ F\circ k_R\ \varphi(R,X_{R-});\ t < R < \zeta \}$$

$$= *P^x\{ E^{X(t)}[Z_R\ g\circ k_R\ \varphi(t+R,X_{R-});\ 0 < R < \zeta]\ h;\ t < R \}$$

$$= *P^x\{ E^{X(t)}[f(X_{R-})\ g\circ k_R\ \varphi(t+R,X_{R-});\ 0 < R < \zeta]\ h;\ t < R \}$$

$$= *P^x\{ f(X_{R-})\ F\circ k_R\ \varphi(R,X_{R-});\ t < R < \zeta \}\ ,$$

and letting $t \downarrow 0$ we obtain (5.8) for $F \in b\,\mathcal{F}^0$ since $*P^x(R = 0) = 0$. This extends immediately to $F \in b\,\mathcal{F}^*$.

We now come to the analog of lemma 4.1. It is the key result.

(5.9) LEMMA. *Let* $F \in b\,\mathcal{F}^*$, $\varphi \in b(\mathcal{B}^+ \times \mathcal{E})$, *and* $Y \in b\,\mathcal{H}^g$. *Then for each* $x \in E$,

(5.10) $$*P^x[F\circ k_R\ \varphi(R,X_{R-})\ Y_R;\ 0 < \tau < R < \zeta]$$
$$= *P^x[P^{x,R,X_{R-}}(F|\tau > 0)\ \varphi(R,X_{R-})\ Y_R;\ 0 < \tau < R < \zeta].$$

PROOF. Recalling (5.6), the left side of (5.10) is equal to

$$*P^x[F\circ k_R\ \varphi(R,X_{R-})\ Y_R;\ Z_R = \infty,\ T < R,\ R < \zeta].$$

But $Y_t\, 1_{\{Z_t = \infty\}}$ is in $b\,\mathcal{H}^g$ and so if f corresponds to it as in (5.7) and (5.8) this last displayed expression may be written

$$*P^x[F\circ k_R\ \varphi(R,X_{R-})\ f(X_{R-});\ \{T < \zeta\}\circ k_R,\ R < \zeta \}$$

since $\{T < R\} = \{T < \zeta\}\circ k_R$.

Now using (3.10) of [4] and reversing the steps this becomes

$$^{*}P^{x}\{\ P^{x,R,X_{R-}}[\ F;\ T<\zeta\ ]\ \varphi(R,X_{R-})\ f(X_{R-});\ R<\zeta\ \}$$

$$=\ ^{*}P^{x}\{\ P^{x,R,X_{R-}}[\ F\mid T<\zeta\ ]\ \varphi(R,X_{R-})\ Y_R;\ 0<\tau<R<\zeta\ \}$$

which yields (5.10), since $\{\tau > 0\} = \{T < \zeta\}$.

We come now to the main theorem. Recall that

$$\Lambda = \{\ G<\tau<\infty,\ D<\infty\ \} = \{\ G<\tau<\infty;\ R\circ\theta_G<\zeta\circ\theta_G\ \}.$$

Also that $\mathcal{F}_{\geq D-}$ is the σ-algebra defined by $Y \in \mathcal{F}_{\geq D-}$ provided $Y \in \mathcal{F}$ and there exists $(Y_t) \in \mathcal{H}^g$ with $Y = Y_D$ on $\{0 < D < \infty\}$.

(5.11) THEOREM. *Let* $F \in b\,\mathcal{F}^*$. *Then on* Λ

$$E^{\mu}[\ F\circ k_R\circ\theta_G \mid \mathcal{F}_G, L, \mathcal{F}_{\geq D-}\ ] = P^{X_G,L,X_{D-}}[\ F \mid \tau>0\ ].$$

Moreover $P^{X_G,L,X_{D-}}(\tau > 0) > 0$ *almost surely on* Λ. *If* $N = \{(x,\ell,y) : P^{x,\ell,y}[X_0 = x,\ \zeta = \ell,\ X_{\zeta-} = y \mid \tau > 0] = 1\}$, *then* $(X_G, L, X_{D-}) \in N$ *almost surely on* Λ.

The proof of (5.11) proceeds from lemma 5.9 exactly as the proof of (4.9) proceeds from (4.1). In fact the argument is somewhat simpler here. In any case we shall not repeat it.

It is instructive to compare (5.11) with (13.7) of [3]. Let $I(\tau) =]G_\tau, D_\tau[$ be the excursion interval straddling τ and $I(T) =]G_T, D_T[$ that straddling T. Recall T is the exact terminal time associated with τ. Since $\{\tau > 0\} = \{T < \zeta\}$ it follows that the law of the excursion straddling τ conditional on $\mathcal{F}_{G(\tau)}$, $\mathcal{F}_{\geq D(\tau)-}$ and $L(\tau)$ on $\{G_\tau < \tau < \infty,\ D_\tau < \infty\}$ is exactly the same as the law of the excursion straddling T conditional on $\mathcal{F}_{G(T)}$, $\mathcal{F}_{\geq D(T)}$, $L(T)$, and $X_{D(T)-}$ on $\{G_T < T < \infty,\ D_T < \infty\}$.

References

1. R. K. GETOOR and M. J. SHARPE. The Markov property at co-optional times. *Z. Wahrscheinlichkeitstheorie verw. Gebiete* **48**, (1979), 201-211.

2. R. K. GETOOR and M. J. SHARPE. Markov properties of a Markov process. *Z. Wahrscheinlichkeitstheorie verw. Gebiete* **55**, (1981), 313-330.

3. R. K. GETOOR and M. J. SHARPE. Excursions of dual processes. *Advances in Mathematics* **45**, (1982), 259-309.

4. R. K. GETOOR and M. J. SHARPE. Two results on dual excursions. *Seminar on Stochastic Processes, 1981*, pp. 31-52. Birkhäuser, Boston, 1981.

5. O. KALLENBERG. Splitting at backward times on regenerative sets. *Ann. Prob.* *9*, (1981), 781-799.

6. B. MAISONNEUVE. On Exit Systems. *Ann. Prob.* *3*, (1975), 399-411.

7. B. MAISONNEUVE. On the structure of certain excursions of a Markov Process. *Z. Wahrscheinlichkeitstheorie verw. Gebiete 47*, (1979), 61-67.

8. P. A. MEYER, R. T. SMYTHE, and J. B. WALSH. Birth and death of Markov processes. *Proc. Sixth Berkeley Symp. Math. Statist. Prob. Vol. III*, 295-306. University of California Press, Berkeley, 1971.

9. M. J. SHARPE. *General Theory of Markov Processes*. Forthcoming book.

R. K. GETOOR
Department of Mathematics
University of California - San Diego
La Jolla, CA 92093

Seminar on Stochastic Processes, 1982
Birkhäuser, Boston, 1983

IDENTIFYING MARKOV PROCESSES UP TO TIME CHANGE*

by

JOSEPH GLOVER

0. Introduction

The intertwining of Markov processes and potential theory has been apparent at least since Hunt's fundamental trilogy on these subjects and was certainly evident even before then. The relationships between these two subjects have been investigated vigorously and profitably since then, and we intend to add to this study here. The central object of interest in potential theory is the cone of excessive functions, a positive cone of functions satisfying various *axioms* or *principles* of potential theory (see for example [7] and the references). The best known is the cone of superharmonic functions in R^3 consisting of positive constants together with functions of the form $\int |x-y|^{-1}\ \mu(dy)$, where μ is a positive measure: this arises in Newtonian potential theory, and today's axiomatic approach owes a great deal to abstraction and generalization of properties of this particular cone of functions. Each reasonable Markov process $(X(t), P^x)$ on E has an associated cone of excessive functions $S(X)$ which can be obtained in an analytic manner from the semigroup $P(t)$: throw a positive function $f(x)$ into the cone if $P(t)f(x) \leq f(x)$ for all positive t and if $P(t)f(x)$ increases to $f(x)$ as

*Research supported in part by NSF Grant MCS-8002659.

t decreases to zero. Such a cone may contain only constant functions. We restrict our detailed discussion to transient processes (see (1.1) for a definition) so that the excessive functions separate points in the state space E. Later in this section, we discuss the non-transient case briefly.

One can look for more direct connections between the sample paths of the process X(t) and its cone of excessive functions, and the first important result in this direction is the Balayage Theorem of Hunt ([2], III-6.12; [8], (12.9)).

(0.1) THEOREM. *If* $X(t)$ *is a transient right process,* $f \in S(X)$, *and* A *is a Borel set, let* $\mathcal{U} = \{u \in S(X): u \geq f \text{ on } A\}$. *If* $f_A = \inf\{u : u \in \mathcal{U}\}$, *then* $P_A f \leq f_A$ *and* $P_A f(x) = f_A(x)$ *except perhaps for those points* x *in* $E \cap (A - A^r)$.

(Here, $P_A f(x) = E^x[f(X(T(A))); T(A) < \infty]$, where $T(A) = \inf\{t > 0: X(t) \in A\}$, and A^r is the collection of regular points of A.) Thus Hunt gave a first step from the potential theory to the process X(t): how to compute the distribution of X(T(A)). Two more important results complete the basic linkage between potential theory and sample paths. Dynkin's theorem ([2], II-5.1) tells how to determine the cone of excessive functions given the hitting distributions. We concentrate our energies on the theorem of Blumenthal, Getoor and McKean ([3],[4]) and its generalizations and spin-offs.

(0.2) THEOREM. *Let* $(X(t), P^x)$ *and* $(Y(t), Q^x)$ *be two standard processes on* $(E(\Delta), \mathcal{E}(\Delta))$ *so that* $P_K(x,\cdot) = Q_K(x,\cdot)$ *for all compact subsets* K *of* $E(\Delta)$. *There is a continuous additive functional* $A(t)$ *of* $X(t)$ *which is strictly increasing and finite on* $[0,\zeta)$ *so that if* $T(t)$ *is the right continuous inverse of* $A(t)$, *then* $(X(T(t)), P^x)$ *and* $(Y(t), Q^x)$ *have the same joint distributions.*

The proof given in [2] applies to right processes as well. These three results seem to completely specify the connections between sample paths and hitting distributions. And yet, in demanding $P_K(x,\cdot)$ in (0.2), we are requiring a prodigious amount of information. One might hope to require different information or less information in order to obtain the conclusion of (0.2). Since last exit times of Markov processes have come to rival in importance first hitting times in recent years, we asked if one could replace the hypothesis of equality of the first hitting distributions with the hypothesis that the last exit distributions are the same and still obtain the conclusion of (0.2). The answer is yes if $X(t)$ and $Y(t)$ are transient Hunt processes [13]:

(0.3) THEOREM. *Let* $(X(t), P^x)$ *and* $(Y(t), Q^x)$ *be transient Hunt processes on* $(E(\Delta), \mathcal{E}(\Delta))$. *Let* L_K *be the last time a process visits a set* K. *Assume that* $P^x[f(X(L_K-)); L_K > 0] = Q^x[f(Y(L_K-)); L_K > 0]$ *for all* x *in* E, *for all bounded continuous functions* f *and for all compact sets* K *in* E. *There is a continuous additive functional* $A(t)$ *of* $X(t)$ *which is strictly increasing and finite on* $[0,\zeta)$ *so that if* $T(t)$ *is the right continuous inverse of* $A(t)$, *then* $(X(T(t)), P^x)$ *and* $(Y(t), Q^x)$ *have the same joint distributions.*

The experienced Markovologist may immediately wonder whether (0.3) can be obtained from (0.2) "merely" by applying time reversal to the result given in (0.2). This does not seem to be the case, and it is interesting to examine what theorem is produced by applying a time reversal argument to the result in (0.2). If we compare (0.2) and (0.3) from the point of view of potential theory, (0.3) seems to require "less information" than (0.2). From Hunt's Balayage theorem and Dynkin's theorem, we see that requiring all of the first hitting distributions is equivalent to requiring the whole cone of excessive functions. Thus the potential theory content of the Blumenthal-Getoor-McKean theorem is that there is at most one right process (up to time change) associated to a cone of excessive functions. Under mild hypotheses (such as duality), the functions in (0.3) can be written as

$$P^x[f(X(L_K-));\ L_K > 0] = Uf\pi_K = \int u(x,y)f(y)\pi_K(dy), \tag{0.4}$$

where $u(x,y)$ is the appropriately regularized potential density of $X(t)$, and π_K is the equilibrium measure of K. Thus the potential theoretic content of (0.3) is the following:

(0.5) THEOREM. *Suppose* $X(t)$ *and* $Y(t)$ *are two transient Hunt processes, each possessing a dual (or satisfying some other hypothesis to ensure the representation* (0.4)). *Suppose* $X(t)$ (*resp.* $Y(t)$) *has potential kernel* U *and equilibrium measures* π_K (*resp. potential kernel* V *and equilibrium measures* γ_K). *If* $Uf\pi_K = Vf\gamma_K$ *for all bounded functions* f *on* E *and for all compact sets* K *contained in* E, *then the class of excessive functions for* X *coincides with the class of excessive functions for* Y; *i.e.* X *and* Y *have the same potential theories.*

Thus we have produced a subcollection of excessive functions which determines the whole cone of excessive functions. From the point of view of the processes, however, it seems that both (0.2) and (0.3) require roughly the "same amount" of information. Namely, for each compact set K, we need a kernel giving either the first hitting or last exit distributions of the process from K.

One is naturally led to wonder what is the "lowest common denominator" of the hypotheses in (0.2) and (0.3). What information do the first hitting distributions and last exit distributions of a set K have in common? They both tell us with what probability K is hit!! In fact, we proved that if X(t) and Y(t) are two transient Hunt processes satisfying the hypothesis of absolute continuity with the same *hitting probabilities* ($P_K 1(x) = Q_K 1(x)$ for all K, for all x), then the conclusions of (0.2) and (0.3) remain true [12]. We show in section 1 that if X(t) and Y(t) are two *transient right processes* satisfying the hypothesis of absolute continuity with the same hitting probabilities, then the conclusion of (0.2) remains true. The arguments in section 1 are similar to and are modelled on those given in [12] for Hunt processes, but incorporate certain delicate compactification arguments. We have chosen to give some arguments in detail, since some of the modifications necessary may not be obvious to those unfamiliar with the arcane delights of compactifications. We "avoided" the use of compactifications in [12] by using the fact that $P^x[f(X(L_K)); L_K > 0]$ can always be represented as a potential of a measure on E if K is compact and X(t) is a transient Hunt process satisfying the hypothesis of absolute continuity. However, we did use compactifications in [14] to prove this representation, so compactifications had already entered the scene.

It is worth pointing out that this result adds something interesting to potential theory as well as Markov process theory. Since two processes as described above are time changes of one another if and only if they have the same hitting probabilities, they must then have the same potential theories. ($P^x(T(K) < \infty)$ is called the *réduite* of 1 on the set K in potential theory, so this shows that if the réduites of 1 on sets are the same, the cones of excessive functions are identical.)

One other point of interest in the extension of (0.2) is that we do not need to assume the topologies on $E(\Delta)$ are the same for both processes. However, we do need to assume that the Borel fields of these topologies are the same.

One can look for other "small" collections of excessive functions which characterize the process up to time change. In section 2, we discuss to what extent the process is determined by its jump probabilities. That is, suppose the probability that a Hunt process $X(t)$ has a jump from K to L is the same as the probability that a Hunt process $Y(t)$ has a jump from K to L for all sets K and L in E. Then, roughly speaking, $X(t)$ and $Y(t)$ are time changes of one another on the support of their Lévy systems (modulo polar sets): see (2.2) for a precise statement.

It should be possible to drop the hypothesis of absolute continuity from all of the *results* we state, but the reader will see that our *methods* depend crucially on it, and so new methods would be needed for such an extension.

Finally, we discuss the extension of the result in section 1 to the case where $X(t)$ and $Y(t)$ need not be transient processes. We shall not prove this extension, but merely indicate methods. Before doing so, we comment on a subtle point in the hypothesis of (0.2). The hypothesis that $P_K(x,\cdot) = Q_K(x,\cdot)$ must be carefully interpreted. In the transient case, it suffices to assume that K is compact and contained in E, so that both measures are measures on E. However, this does *not* suffice in the case of general right (or even Hunt) processes. For example, let $E = \{x\}$, so $E(\Delta) = \{x,\Delta\}$. The process $X(t)$, starting at x, sits there forever. The process $Y(t)$, starting at x, sits there an exponential amount of time and then jumps to the cemetery Δ. Since $P_{\{x\}}(x,\{x\}) = 1 = Q_{\{x\}}(x,\{x\})$, $P_K = Q_K$ for all compact sets K contained in E. Every continuous additive functional of $X(t)$ is of the form ct, and every continuous additive functional of $Y(t)$ is of the form $d(t \wedge \zeta)$. It is easy to see that the processes are not time changes of one another as described in (0.2). In fact, in the general case of the Blumenthal-Getoor-McKean theorem, one needs to assume that $P_K(x,\cdot) = Q_K(x,\cdot)$ are measures on $E(\Delta)$. (This rules out the example above. There, $P_\Delta(x,\cdot) = 0$, and $Q_\Delta(x,\cdot) = \varepsilon_\Delta(\cdot)$.) Thus Δ is considered as just another trap in the state space (in contrast with much of Markov process theory, where Δ is ignored as much as possible as a "cemetery").

Let $X = (\Omega, F, F_t, X_t, \theta_t, P^x)$ and $Y = (\Omega, G, G_t, Y_t, \theta_t, Q^x)$ be two right processes on a Lusin space $(E(\Delta), \mathcal{E}(\Delta))$ which satisfy the hypothesis of absolute continuity. We no longer require them to be transient, and it is appropriate in this case to assume that $P^x(T(K) < T(L)) = Q^x(T(K) < T(L))$ for all compact subsets K and L of $E(\Delta)$. Then $X(t)$ and $Y(t)$ have the same traps, for z is a trap for $X(t)$ and $Y(t)$ if and only if $P^z(T(K) < T(\{\Delta\})) = 0 = Q^z(T(K) < T(\{\Delta\}))$ for all compact sets K contained in $\{z\}^c$. Let T denote the set of traps of the two processes: T^c is finely open and finely closed for both processes. Suppose K is a nonpolar "exit set" in T^c so that $P^x(T(K^c) < \infty) = 1$ and $Q^x(T(K^c) < \infty) = 1$. Then $P^x(T(L) < T(K^c)) = Q^x(T(L) < T(K^c))$ means that the process $X(t)$ killed at $T(K^c)$ has the same hitting probabilities as $Y(t)$ killed at $T(K^c)$. The results of section 1 will guarantee that these two processes are identical after a time change. Then one must show that $X(t)$ and $Y(t)$ are time changes of one another by "piecing together" these results on exit sets like K above. This procedure is carried out in the Blumenthal-Getoor-McKean theorem, and we refer the reader to V-5 in [2] for a guide to the procedure. Just as $P^x(T(K) < \infty)$ has a significance as a réduite of 1 on K in potential theory, $P^x(T(L) < T(K))$ also has a significance: Chung and Getoor [5] have identified functions of this form as *condensor potentials* using various hypotheses.

Let us illustrate the idea of the proof of section 1 by examining the simplest case where X(t) and Y(t) are transient right processes on a finite state space $E = \{1,2,3,\ldots,n\}$. In this case, X(t) and Y(t) must be Hunt processes and have duals. Moreover, $P^x(T(K) < \infty) = P^x(L_K > 0) = U\pi_K(x)$ and $Q^x(T(K) < \infty) = V\gamma_K(x)$. Taking $K = \{y\}$, we observe that $\pi_K(dz) = c(y)\varepsilon_{\{y\}}(dz)$ and $\gamma_K(dz) = d(y)\varepsilon_{\{y\}}(dz)$, where $c(y) > 0$ and $d(y) > 0$. Thus $u(x,y)c(y) = v(x,y)d(y)$. If we set $A(t) = \int_0^t (c/d)(X(s))ds$ and $T(t) = \inf\{s\colon A(s) > t\}$, then X(T(t)) has the potential kernel $u(x,y)c(y)/d(y) = v(x,y)$. It follows from V-5.10 in [2] that $(X(T(t)), P^x)$ has the same distribution as $(Y(t), Q^x)$.

The standard notation of Markov processes is used throughout: see, for example, [2] and [8]. If K is any metric space, $bC(K)^+$ denotes the bounded positive continuous functions on K. The indicator of a set K is denoted by $\chi(K)$ or $\chi(K)(x)$.

1. The Time Change Theorem for Transient Right Processes

All of the hypotheses for this section are contained in this first paragraph. Let $E(\Delta)$ be a set of points containing a point Δ (which plays the role of a cemetery for the Markov processes), and let τ and σ be two topologies on $E(\Delta)$ so that $(E(\Delta), \tau)$ and $(E(\Delta), \sigma)$ are both Lusin topological spaces with the same Borel field $\mathcal{E}(\Delta)$. Let $\tilde{X} = (\Omega, \tilde{F}, \tilde{F}_t, \tilde{X}_t, \tilde{\theta}_t, \tilde{P}^x)$ be a right process on $(E(\Delta), \tau)$ with semigroup $\tilde{P}(t)$ and resolvent $\tilde{U}^a$ [8]. Let $\tilde{Y} = (W, \tilde{G}, \tilde{G}_t, \tilde{Y}_t, \tilde{\theta}_t, \tilde{Q}^x)$ be a right process on $(E(\Delta), \sigma)$ with semigroup $\tilde{Q}(t)$ and resolvent $\tilde{V}^a$. We assume that there exist two bounded Borel functions h^X and h^Y which are strictly positive on $E = E(\Delta) - \{\Delta\}$ so that

$$(1.1) \qquad \tilde{U}h^X \le 1 \quad \text{and} \quad \tilde{V}h^Y \le 1.$$

Processes satisfying this assumption are called *transient*. We assume that η and ρ are two *reference probability measures* for $\tilde{X}$ and $\tilde{Y}$, respectively: $\tilde{U}^a(x,\cdot) << \eta$ and $\tilde{V}^a(x,\cdot) << \rho$ for all nonnegative a. Finally, we assume that $\tilde{X}$ and $\tilde{Y}$ have the same hitting probabilities:

$$\tilde{P}^x(T(K) < \infty) = \tilde{Q}^x(T(K) < \infty) \quad \text{for all sets } K \in \mathcal{E}. \tag{1.2}$$

If $\tau = \sigma$, it suffices to assume (1.2) holds for all compact sets $K \subset E$. We shall prove:

THEOREM. *Let $\tilde{X}$ and $\tilde{Y}$ be two right processes as described in the paragraph above. Then there is a continuous additive functional $H(t)$ of $\tilde{Y}$ which is strictly increasing and finite up to the lifetime of $\tilde{Y}$ so that if we set $\Xi(t) = \inf\{s: H(s) > t\}$, then $(\tilde{X}(t), \tilde{P}^x)$ has the same law as that of the process $(\tilde{Y}(\Xi(t)), \tilde{Q}^x)$.*

We use two important consequences of hypotheses (1.1) and (1.2) over and over again in this section without explicitly mentioning them each time. First, (1.2) implies that

(1.3) $\tilde{X}$ *and* $\tilde{Y}$ *have the same polar sets.*

Second, (1.1) implies that there is a sequence $T(X,n)$ (resp. $T(Y,n)$) of sets in $E(\Delta)$ which are increasing and finely open for $\tilde{X}$ (resp. $\tilde{Y}$) and so that $\sup\{t: \tilde{X}(t) \in T(X,n)\} < \infty$ a.s. (resp. $\sup\{t: \tilde{Y}(t) \in T(Y,n)\} < \infty$ a.s.). We shall call a set $G \in \mathcal{E}$ *transient* if $L(G) = L_G = \sup\{t: \tilde{X}(t) \in G\} < \infty$ a.s. and if $L(G) = L_G = \sup\{t: \tilde{Y}(t) \in G\} < \infty$ a.s. Therefore,

(1.4) *if* $K \in E$ *is not polar, there is a transient set* $G \subset K$ *that is not polar.*

Set $A(t) = \int_0^t h^X(\tilde{X}(s))ds$, $B(t) = \int_0^t h^Y(\tilde{Y}(s))ds$, $T(t) = \inf\{s: A(s) > t\}$, $S(t) = \inf\{s: B(s) > t\}$. If we set $F = \tilde{F}$, $F_t = \tilde{F}_{T(t)}$, $X(t) = \tilde{X}(T(t))$, $\theta_t = \tilde{\theta}_{T(t)}$, and $P^x = \tilde{P}^x$, then $X = (\Omega, F, F_t, X_t, \theta_t, P^x)$ is a right process on $(E(\Delta), \tau, E(\Delta))$ with semigroup $P(t)$ and resolvent U^a. If we set $G = \tilde{G}$, $G_t = \tilde{G}_{S(t)}$, $Y(t) = \tilde{Y}(S(t))$, $\theta_t = \tilde{\theta}_{S(t)}$, and $Q^x = \tilde{Q}^x$, then $Y = (W, G, G_t, Y_t, \theta_t, Q^x)$ is a right process on $(E(\Delta), \sigma, E(\Delta))$ with semigroup $Q(t)$ and resolvent V^a. If $\zeta = \inf\{t: X(t) = \Delta\}$ and $z = \inf\{t: Y(t) = \Delta\}$, then $\zeta < \infty$ a.s. and $z < \infty$ a.s. since $U1 = \tilde{U}h^X \leq 1$ and $V1 = \tilde{V}h^Y \leq 1$. Note that X and Y have the same hitting probabilities as $\tilde{X}$ and $\tilde{Y}$, and (1.3) and (1.4) hold for X and Y.

Using time reversal, one can construct left continuous moderate Markov processes $(\hat{X}(t), \hat{P}^x)$ and $(\hat{Y}(t), \hat{Q}^x)$ with moderate Markov semigroups $\hat{P}(t)$ and $\hat{Q}(t)$ and resolvents $\hat{U}^a$ and $\hat{V}^a$ so that $\hat{U}^a(\cdot,x) << \lambda = \eta U$, $\hat{V}^a(\cdot,x) << \xi = \rho V$, $\lambda(f \cdot U^a g) = \lambda(f\hat{U}^a \cdot g)$, and $\xi(f \cdot V^a g) = \xi(f\hat{V}^a \cdot g)$ for all positive Borel functions f and g ([6], [16], [18]). (Here, $\lambda(f)$ means $\int f(x)\lambda(dx)$, and coresolvents act on functions on the left — see Chapter VI of [2]). As in Chapter VI of [2], for each $a \geq 0$, we may choose potential densities $u^a(x,y)$ and $v^a(x,y)$ in $E(\Delta)\times E(\Delta)$ having the properties:

(i) $x \to u^a(x,y)$ is a-excessive for (U^a).
$x \to v^a(x,y)$ is a-excessive for (V^a).

(ii) $y \to u^a(x,y)$ is a-excessive for $(\hat{U}^a)$.
$y \to v^a(x,y)$ is a-excessive for $(\hat{V}^a)$.

(iii) $U^a f(x) = \int u^a(x,y)\, f(y)\, \lambda(dy), \quad f \in \mathcal{E}(\Delta)^+.$

$V^a f(x) = \int v^a(x,y)\, f(y)\, \xi(dy), \quad f \in \mathcal{E}(\Delta)^+.$

(iv) $f\hat{U}^a(y) = \int f(x)\, u^a(x,y)\, \lambda(dx), \quad f \in \mathcal{E}(\Delta)^+.$

$f\hat{V}^a(y) = \int f(x)\, v^a(x,y)\, \xi(dx), \quad f \in \mathcal{E}(\Delta)^+.$

The moderate Markov duals above lack some of the nice properties of right processes, in general. For example, they may not have right continuous strong Markov versions on $E(\Delta)$, and they may not be normal. Compactification techniques have proved to be useful tools in dealing with such processes in the past, and we shall find them useful again here. In fact, this is why the initial topologies τ and σ may differ: we are going to replace them with more "natural" topologies. In [11], we constructed a compact metric space $E(X)$ with Borel field $\mathcal{E}(X)$ so that U^a and $\hat{U}^a$ extend to be *Ray resolvents* on $E(X)$ (which extensions we again denote by U^a and $\hat{U}^a$), $E \in \mathcal{E}(X)$, and $E(\Delta)$ is dense in $E(X)$. The procedure is given in [11] in detail, and a succinct summary of the results of [11] and several complements to these results are given in section 2 of [14]. Therefore, we shall refer to these two articles whenever we use these results and avoid repeating details. Let $(E(Y), \mathcal{E}(Y))$ be the analogous compactification for $Y(t)$ and $\hat{Y}(t)$. Set

$$D(X) = \{x \in E(X)\colon \lim_{a\to\infty} af\hat{U}^a(x) = f(x) \text{ for all } f \in bC(E(X))^+\}$$

$$D(Y) = \{x \in E(Y)\colon \lim_{a\to\infty} af\hat{V}^a(x) = f(x) \text{ for all } f \in bC(E(Y))^+\}$$

$$C(X) = D(X) \cap E$$

$$C(Y) = D(Y) \cap E$$

$$B(X) = \{x \in E(X)\colon \hat{U}^a(\cdot,x) << \lambda \text{ for all } a\}$$

$$B(Y) = \{x \in E(Y)\colon \hat{V}^a(\cdot,x) << \xi \text{ for all } a\}.$$

(Note: in [14], D(X) was called $\hat{D}$ and C(X) was called $\hat{C}$; we drop the circumflexes here.) Then $\hat{U}^a$ (resp. $\hat{V}^a$) restricted to D(X) (resp. D(Y)) is the resolvent of a right process on D(X) (resp. D(Y)). If x is in C(X) (resp. C(Y)), then $(\hat{X}(t+), \hat{P}^x)$ (resp. $(\hat{Y}(t+), \hat{Q}^x)$) is a realization of the right process (where the right limit is taken in the topology of E(X) (resp. E(Y))). It will be important to recall that $\lambda((B(X) \cap E)^c) = \xi((B(Y) \cap E)^c) = 0$ (Lemma (6.2), [11]). Also recall that $E(X) - D(X)$ is semipolar for $\hat{X}(t)$ and $E(Y) - D(Y)$ is semipolar for $\hat{Y}(t)$: $\hat{P}^x(\hat{X}(t) \in E(X) - D(X)$ for some $t \geq 0) = 0$ and $\hat{P}^x(\hat{X}(t-) \in E(X) - D(X)$ uncountably often$) = 0$. Moreover, $\lambda((B(X) \cap C(X))^c) = 0$ and $\xi((B(Y) \cap C(Y))^c) = 0$. The densities $u^a(x,y)$ and $v^a(x,y)$ may be extended to be densities on $E \times B(X)$ and $E \times B(Y)$, respectively, so that the formulae in (iv) above hold for all y in B(X) (resp. B(Y)) (see the two paragraphs following (6) in [14]).

This compactification was used in [14] to prove that $P^x(T(K) < \infty)$ is the potential of a measure π_K, where $\pi_K(E(X) - E) = 0$ if K is a compact set contained in E and X is a *Hunt* process. In general, π_K may charge $E(X) - E$. Consult (7.2) and (7.3) in [11] for the following result.

(1.5) THEOREM. *Let* A(t) *and* B(t) *be predictable additive functionals of* X(t) *and* Y(t) *with bounded* 1*-potentials. Then*

$$P^x \int f(X(t-))dA(t) = \int u(x,y)\, f(y)\, \nu(dy), \text{ and}$$

$$Q^x \int f(Y(t-))dB(t) = \int v(x,y)\, f(y)\, \mu(dy),$$

where ν *is the Revuz measure of* A(t), *and* μ *is the Revuz measure of* B(t); ν *is concentrated on* B(X), *and* μ *is concentrated on* B(Y). (Here, X(t-) and Y(t-) are the left limits of X(t) and Y(t), taken in the topologies of E(X) and E(Y), respectively.)

(1.6) COROLLARY. *If* K *is a transient set in* E, *then*

$$P^x(T(K) < \infty) = U\pi_K(x) = \int u(x,y)\, \pi_K(dy) \text{ and}$$

$$Q^x(T(K) < \infty) = V\gamma_K(x) = \int v(x,y)\, \gamma_K(dy),$$

where π_K *is concentrated on the closure of* K *in* E(X) *and* γ_K *is concentrated on the closure of* K *in* E(Y). *Moreover,* $\pi_K(B(X)^c) = \gamma_K(B(Y)^c) = 0$.

PROOF. The last statement follows from the last statement in (1.5). Now $P^x(T(K) < \infty) = P^x(L(K) > 0) = P^x(A(\infty))$, where A(t) is the dual predictable projection of the raw (i.e. nonadapted) additive functional $\chi(0 < L(K) \leq t)$. Thus $P^x(A(\infty)) = U\pi_K(x)$ by (1.1). Let J be the indicator of the complement of the closure of K in E(X). Then $P^x \int J(X(t-))dA(t) = P^x[J(X(L(K)-); 0 < L(K)] = 0$, so $\pi_K(J) = 0$. The argument for γ_K is the same. Q.E.D.

(1.7) LEMMA. *There are functions* p^X *and* p^Y *on* E(X) *and* E(Y) *which are strictly positive and bounded on* D(X) *and* D(Y) *so that* $p^X\hat{U}(x) \leq 1$ *for all* x *in* E(X) *and* $p^Y\hat{V}(x) \leq 1$ *for all* x *in* E(Y).

PROOF. Since $\infty > \lambda(U1) = \lambda(1\hat{U})$, $1\hat{U}(x) < \infty$ except on some polar set Γ. Therefore, $\hat{X}(t+)$ restricted to $D(X) - \Gamma$ has $1\hat{U} < \infty$. By Proposition (2.2) of [10], there is a strictly positive bounded function p^X so that $p^X\hat{U} \leq 1$ on $D(X) - \Gamma$, so $p^X\hat{U} \leq 1$ on all of E(X). The argument for $\hat{Y}$ is the same. Q.E.D.

Our aim is to show that there is a function f so that $u(x,y) = v(x,y)f(y)$ a.s. (λ) for each x. To do this, we find it convenient to introduce yet another topology on $E(\Delta)$. The construction of this topology is exactly as in [12], so we do not lavish many words on it.

Since the constant function 2 is excessive for $\hat{U}^a$ restricted to $D(X)$, we can find bounded positive functions (g_n) which are λ-integrable so that $g_n\hat{U}$ increases to 2 on $D(X)$. Set

$$S^+ = \{\sum_{j=1}^{n} (p^X p_j + c_j g_j)\hat{U}(x):\ (p_j) \subset C(E(X))^+,\ c_j \geq 0,\ n \geq 1\}.$$

$$R^+ = \{\sum_{j=1}^{n} a_j p_1^j \wedge \cdots \wedge p_{n_j}^j \wedge c_j:\ (p_m^j) \subset S^+,\ a_j \geq 0,\ c_j \geq 0,\ n_j \geq 1\}.$$

Then S^+ is separable in the uniform norm since $p^X\hat{U} \leq 1$; so $R = R^+ - R^+$ is separable in the uniform norm and also separates points on $C(X)$. Using the identity $(a-b)\wedge(c-d) = (a+d)\wedge(c+b) - (d+b)$, one can check that R is a vector lattice.

Let $F = \prod_{j=1}^{\infty} [0,1]$, let (h^j) be a sequence of functions in R which is dense in R in the uniform norm, and let $\Phi: C(X) \to F$ by setting

$$\Phi(x) = (h^j(x)/||h^j||_\infty)_{j\geq 1}.$$

We take the metric on $C(X)$ induced by a metric on F compatible with the product topology of F and given by

$$d(x,y) = \sum_{j=1}^{\infty} 2^{-j} \frac{|h^j(x) - h^j(y)|}{1 + |h^j(x) - h^j(y)|}.$$

Complete $C(X)$ in this metric to obtain a compact metric space $\bar{C}(X)$. If g is an element of R, let $\bar{g}$ denote the continuous extension of g to $\bar{C}(X)$, and let $\bar{R} = \{\bar{g}:\ g \in R\}$. Now $\bar{R}$ may not contain the constant functions, so we cannot assert that $\bar{R}$ is dense in $C(\bar{C}(X))$. But if $\bar{C}(n,X)$ is the closure in $\bar{C}(X)$ of $C(n,X) = \{x \in C(X): g_n\hat{U}(x) > 1\}$, then $\bar{R}(\bar{C}(n,X))$ (defined to be the restrictions to $\bar{C}(n,X)$ of functions in $\bar{R}$) contains the constant functions on $\bar{C}(n,X)$ and so is dense in $C(\bar{C}(n,X))$ (by the vector-lattice form of the Stone-Weierstrass theorem).

(1.8) PROPOSITION. *To each* f *in* R, *there corresponds a finite signed measure* ν *on* E *so that* $f(x) = \nu\hat{U}(x)$ *for all* x *in* D(X).

PROOF. By construction, every function f in S satisfies the proposition. To complete the proof, we need only show that if $f\hat{U}(x)$ is in S^+ and if $g\hat{U}(x)$ is in S^+, then $F(x) = \min(f\hat{U}(x), g\hat{U}(x))$ satisfies the proposition. Notice that $F(x)$ is excessive for the resolvent $\hat{U}^a$ restricted to D(X). Since $F(\hat{X}(t+))$ is a right continuous supermartingale dominated by the potential $f\hat{U}(\hat{X}(t+))$, there is a predictable additive functional $A(t)$ not charging $\hat{\zeta}$ with $F(x) = \hat{P}^x(A(\infty))$ for all x in D(X). Let ν be the Revuz measure of $A(t)$, defined by setting $\nu(g) = \lim_{a\to\infty} a\hat{P}^\lambda \int e^{-as} g(\hat{X}(s-))dA(s)$ for g in $bE(X)^+$. Since $\hat{P}^\lambda(\hat{X}(s-) \in E(X) - E(\Delta)$ for some $s > 0) = 0$, $\nu(E(X) - E(\Delta)) = 0$. The representation theorem stated in (1.1) applies equally well to the process $\hat{X}(t)$, and we get that $F(x) = \hat{P}^x(A(\infty)) = \nu\hat{U}(x)$ for all x in D(X). We may find a sequence of positive functions (ϕ_k) so that $U\phi_k(x)$ increases to 1 on E. Therefore, $\infty > \lambda(g) = \lim_{k\to\infty} \int g\hat{U}(x)\phi_k(x)\lambda(dx) \geq \lim_{k\to\infty} \int \nu\hat{U}(x)\phi_k(x) \cdot\lambda(dx) = \nu(E)$. Q.E.D.

We now fix n and work with $C(n,X)$. If $\bar{K} \subset \bar{C}(n,X)$ is closed, there is a bounded sequence of functions $(\phi_k) \subset C(\bar{C}(n,X))^+$ so that ϕ_k decreases to $\chi(\bar{K})$. For each k, choose $\bar{h}_k \in \bar{R}$ so that $\sup\{|\bar{h}_k(x) - \phi_k(x)| : x \in \bar{C}(n,X)\} < 1/k^2$. Then $\overline{\bar{h}_k + (1/k^2)g_n\hat{U}(x)}$ is positive and converges boundedly to $\chi(\bar{K})$ on $\bar{C}(n,X)$. Let $K = \bar{K} \cap C(n,X)$. Since $h_k + (1/k^2)g_n\hat{U}(x) \in R$ is of the form $\mu_k\hat{U}(x)$ for all x in $C(X)$, $\mu_k\hat{U}(x)\chi(C(n,X))(x)$ converges boundedly to $\chi(K)(x)$ as k increases to infinity.

Let $L \subset C(n,X)$ be nonpolar, transient, closed in $\bar{C}(n,X)$ and with closure in $\bar{C}(n,Y)$ contained in $C(n,X)$. Then $P^x(T(L) < \infty) = U\pi_L(x)$ (with $\pi_L(E(X) - L) = 0$), and $P^{\mu(k)}(T(L) < \infty)$ converges to $\pi_L(K) = \pi_L(\bar{K})$ by the Lebesgue dominated convergence theorem. Since μ_k is a measure on E and $L \subset E$, it makes sense to talk about $Q^{\mu(k)}(T(L) < \infty)$, and $P^{\mu(k)}(T(L) < \infty) = Q^{\mu(k)}(T(L) < \infty)$. If we use the notation $\mu_k U\pi_L$ to denote $\int U\pi_L(x)\mu_k(dx)$, then we have that $\mu_k U\pi_L = \mu_k V\gamma_L$ converges to $\pi_L(K)$. Recall that γ_L is concentrated on $C(n,X)$ by our choice of L. We must be cautious: there are three topologies in use simultaneously!!

(1.9) LEMMA. *$\mu_k\hat{V}(x)$ is well-defined on $C(n,X)$ except possibly on a polar set Γ_k.*

PROOF. We need only show that $\Gamma_k = \{|\mu_k|\hat{V} = \infty\} \cap E$ is a polar set. Let L be a transient set contained in Γ_k which is closed in $E(Y)$. Then $|\mu_k|V\gamma_L = |\mu_k|U\pi_L < \infty$ since $|\mu_k|\hat{U}$ is bounded. But if γ_L is not zero, then $|\mu_k|V\gamma_L = \infty$ since $\gamma_L(\Gamma_k^c) = 0$. Thus, γ_L is zero, so L is polar. Therefore, $\mu_k\hat{V}(x) = \mu_k^+\hat{V}(x) - \mu_k^-\hat{V}(x)$ is well-defined except perhaps on Γ_k.

Q.E.D.

We set $\Gamma = \bigcup_k \Gamma_k$. Note that Γ depends on the sequence (μ_k). As mentioned above, we are trying to show that $u(x,y) = v(x,y)f(y)$ a.s. (λ) for each x. Since λ is a measure on $C(X)$, we are more concerned at the moment with the behavior of $\mu_k\hat{V}(x)$ on $C(n,X)$ than with the behavior on $E(Y)$.

(1.10) PROPOSITION. $\mu_k\hat{V}(x)$ *converges boundedly on* $C(n,X)$ - {a polar set depending on (μ_k)} *to a function* $f(K)$ *supported on* $K \cup$ {a polar set depending on (μ_k)}. *Moreover,* $\{\mu_k\hat{V}(x) < 0\} \cap (C(n,X) - \Gamma)$ *is polar.*

PROOF. Let L be any transient subset of $G = \{\mu_k\hat{V}(x) < 0\} \cap (C(n,X) - \Gamma)$ which is *closed* in E(Y). Then γ_L is concentrated on L, while the construction of the (μ_k) guarantees that $0 \leq \mu_k U\pi_L = \mu_k V\gamma_L \leq 0$. Thus $\pi_L = \gamma_L = 0$, so L must be polar, and this proves the last sentence. (Remark: We have implicitly used (1.3) and some well-known facts here. If G is not polar (for X and Y by (1.3)), then there is a *compact* subset of G which is not polar — and the compactness can be relative to the topology of E(X) or the topology of E(Y).) Now let

$$H(k,x) = \mu_k\hat{V}(x) + (2/k^2)g_n\hat{V}(x) - \mu_{k+1}\hat{V}(x)$$

$$I(k,x) = \mu_k\hat{U}(x) + (2/k^2)g_n\hat{U}(x) - \mu_{k+1}\hat{U}(x)$$

$$G(k) = \{x\colon H(k,x) < 0\} \cap (C(n,X) - \Gamma).$$

If G(k) is nonempty, choose a transient subset $L \subset G(k)$ which is closed in E(Y) so that γ_L is concentrated on L. Then $\gamma_L(H(k,x)) < 0$. Then $\pi_L(I(k,x)) < 0$. But our construction of (μ_k) guarantees that $I(k,x) > 0$ on $C(n,X)$. Therefore, $\gamma_L = \pi_L = 0$, and G(k) is polar. So if we set $E(n,X) = (\bigcup_k G(k) \cup \Gamma)^c \cap C(n,X)$, $H(k,x) \geq 0$ on $E(n,X)$, and we conclude that $\mu_k\hat{V}(x)$ converges to some function $f(K)(x)$ on $E(n,X)$. Moreover, $0 \leq \mu_k\hat{V}(x) \leq \mu_1\hat{V}(x) + 2\left(\sum k^{-2}\right)g_n\hat{V}(x)$ on $C(n,X)$. To examine the support of f(K), let L be a subset of $C(n,X)-K$ which is *closed* in E(X). Then $\gamma_L(f(K)) = \lim_{k\to\infty} \mu_k V\gamma_L = \lim_{k\to\infty} \mu_k U\pi_L = \pi_L(K) = 0$. Q.E.D.

(1.11) PROPOSITION. *If* $\bar{J} \subset \bar{K} \subset \bar{C}(n,X)$ *are compact sets in the topology of* E(X), *then* $\{x\colon f(K)(x) \neq f(J)(x)\} \cap J$ *is a polar set.*

PROOF. If $M = \{f(K) > f(J)\} \cap J$ is not polar, let N be any nonpolar subset of M which is closed in the topology of E(Y). Then there is a nonpolar set $L \subset N$ which is closed in the topology of E(X). Then π_L is concentrated on L, so that $\gamma_L(f(J)) = \pi_L(J) = \pi_L(K) = \gamma_L(f(K))$. Thus L must be polar. Therefore, M is polar, and the case $\{f(K) < f(J)\} \cap J$ is the same. Q.E.D.

Thus there is a function f(x) defined up to polar sets so that $\pi_L(\bar{K}) = \pi_L(K) = \gamma_L(f\ \chi(K)) = \gamma_L(f\ \chi(\bar{K}))$ for sets $\bar{K}$ which are compact in $\bar{C}(n,X)$ (for some n) and for sets $L \subset C(n,X)$ which are closed in $\bar{C}(n,X)$ and which have closure in E(Y) contained in C(n,X). By inner regularity of the measures on $\bar{C}(n,X)$, it follows that $\pi_L(B) = \gamma_L(f\ \chi(B))$ for all Borel sets $B \subset \bar{C}(n,X)$ and hence for all Borel sets $\dot{B} \subset E(X)$. Therefore, if L is as above, $U\pi_L(x) = Uf\gamma_L(x) = V\gamma_L(x)$, and it follows that

$$\Lambda(n,x) = \{y \in C(X): v(x,y) \neq u(x,y)\ f(y)\}$$

is a polar set. It is easy to check that $\{f = 0\}$ is polar, so f^{-1} makes sense. Since λ is concentrated on C(X), $Ug(x) = Uf(f^{-1}g)(x) = \int v(x,y)f^{-1}(y)g(y)\lambda(dy)$. And this shows that the cone of excessive functions for X, S(X), is contained in the cone of excessive functions for Y, S(Y).

Reversing the roles of X(t) and Y(t), we see that the same arguments show S(Y) is contained in S(X), so S(X) = S(Y). By Hunt's balayage theorem, X(t) and Y(t) have the same hitting distributions: $P_K(x,\cdot) = Q_K(x,\cdot)$. The Blumenthal-Getoor-McKean theorem now applies to give the final result. One can avoid using the Blumenthal-Getoor-McKean theorem, if desired, as follows. The function U1(x) is a bounded excessive function for the process Y(t) and is the potential of a strictly increasing continuous additive functional $Z(t)$ of $Y(t)$. If $S(t)$ denotes the right continuous inverse of $Z(t)$, it is easy to check that if $V_Z^a g(x)$ is defined to be $Q^x \int e^{-at} g(Y(S(t)))dt$, then $V_Z = U$. This implies that $V_Z^a = U^a$ for all positive a. Q.E.D.

2. Processes with Identical Jump Probabilities

In this section, we assume $X = (\Omega, F, F_t, X_t, \theta_t, P^x)$ and $Y = (\Omega, G, G_t, Y_t, \theta_t, Q^x)$ are transient *Hunt* processes on a Lusin topological space $E(\Delta)$ with Borel field $E(\Delta)$. Assume that both $X(t)$ and $Y(t)$ satisfy the hypothesis of absolute continuity, and let $u(x,y)$ and $v(x,y)$ be their potential densities as in section 1.

(2.1) THEOREM. *Let* $A(t)$ *and* $B(t)$ *be continuous additive functionals of* $X(t)$ *and* $Y(t)$ *with finite* 1*-potentials. Then* $P^x(A(\infty)) = U\mu(x)$ *and* $Q^x(B(\infty)) = V\nu(x)$, *where* μ *and* ν *are measures on* (E, E).

PROOF. By Theorem (7.3) of [11], $P^x(A(\infty)) = U\mu(x)$, where μ is a measure on the compactification of $E(\Delta)$ discussed in section 1. There, μ is the Revuz measure of $A(t)$, defined by setting

$$\mu(g) = \lim_{a\to\infty} aP^\lambda \int e^{-at} g(X^*(t-))\, dA(t)$$

(where $X^*(t-)$ is the left limit of $X(t)$ taken in the topology of the compactification). Since $A(t)$ is continuous, $dA(t)$ does not charge the discontinuities of $X(t)$, so

$$\mu(g) = \lim_{a\to\infty} aP^\lambda \int e^{-at} g(X(t))\, dA(t),$$

and this measure is concentrated on $E(\Delta)$ since the state space of $X(t)$ is $E(\Delta)$. Q.E.D.

Define:

$$J(X,R,S) = \chi(\{X(t-) \in R,\ X(t) \in S \text{ for some } t > 0\})$$

$$J(Y,R,S) = \chi(\{Y(t-) \in R,\ Y(t) \in S \text{ for some } t > 0\}).$$

In order to state the main theorem of this section, we need to recall a few facts from the construction of Lévy systems. We follow the argument of Benveniste and Jacod [1], but the results are a bit simpler here than in their situation. We are dealing with Hunt processes, while they discuss general right processes. They construct strictly positive functions g and h so that the additive functionals

$$C(t) = \sum_{s \le t} \chi(\{X(s-) \ne X(s)\}) g(X_{s-}, X_s)$$

and

$$D(t) = \sum_{s \le t} \chi(\{Y(s-) \ne Y(s)\}) h(Y_{s-}, Y_s)$$

have finite 1-potentials. Let $Z(s)$ be a bounded predictable process, and let $F(x,y)$ be a bounded measurable function on $E \times E$. Then

$$P^x \sum_s Z(s) F(X(s-), X(s)) g(X_{s-}, X_s) \chi(\{X(s-) \ne X(s)\}) = P^x \int Z(s)\, dA^F(s)$$

and

$$Q^x \sum_s Z(s) F(Y(s-), Y(s)) h(Y_{s-}, Y_s) \chi(\{Y(s-) \ne Y(s)\}) = Q^x \int Z(s)\, dB^F(s),$$

where $A^F(s)$ and $B^F(s)$ are the dual predictable projections of the processes $\int_0^s F(X(t-), X(t))\, dC(t)$ and $\int_0^s F(Y(t-), Y(t))\, dD(t)$. Since $C(t)$ and $D(t)$ are purely discontinuous, quasi-left-continuous additive functionals, $A^F(s)$ and $B^F(s)$ are continuous additive functionals. If $A(s)$ and $B(s)$ are the dual predictable projections of $C(s)$ and $D(s)$, then $dA^F(s) \ll dA(s)$ and $dB^F(s) \ll dB(s)$. By Motoo's theorem, $dA^F(s) = p(X(s))dA(s)$ and $dB^F(s) = q(Y(s))dB(s)$ for some functions p and q depending on F. Using a standard argument [9], one can produce kernels $K(x,dy,dw)$ and $L(x,dy,dw)$ so that

$$P^x \sum_s Z(s)F(X(s-),X(s))g(X_{s-},X_s)\chi(\{X(s-) \neq X(s)\}) = P^x \int Z(s)KF(X(s))dA(s)$$

and

$$Q^x \sum_s Z(s)F(Y(s-),Y(s))h(Y_{s-},Y_s)\chi(\{Y(s-) \neq Y(s)\}) = Q^x \int Z(s)LF(Y(s))dB(s).$$

Define $S(t) = \inf\{s: A(s) > t\}$ and $T(t) = \inf\{s: B(s) > t\}$. Then $\Gamma = \{x: P^x(S(0) = 0) = 1\}$ and $\Lambda = \{x: Q^x(T(0) = 0) = 1\}$ are the fine supports of the continuous additive functionals $A(t)$ and $B(t)$, and the processes $X_S = (X(S(t)), (P^x)_{x \in \Gamma})$ and $Y_T = (Y(T(t)), (Q^x)_{x \in \Lambda})$ are right processes. The potential kernels of X_S and Y_T are $(U_A(x,\cdot))_{x \in \Gamma}$ and $(V_B(x,\cdot))_{x \in \Lambda}$, respectively.

(2.2) THEOREM. *If* $P^x(J(X,R,S)) = Q^x(J(Y,R,S))$ *for all* x *and for all pairs of compact sets* R *and* S *in* E*, then*

(i) *there are continuous additive functionals* $G(t)$ *and* $H(t)$ *of* $X(t)$ *and* $Y(t)$ *so that* $U_A = V_H$ *and* $V_B = U_G$ *on* E.

(ii) *If* $\Gamma = \Lambda = E$*, then* $X(t)$ *and* $Y(t)$ *are time changes of one another.*

PROOF. Let ν be the Revuz measure of $A(t)$, and let μ be the Revuz measure of $B(t)$. If G is measurable and $(\mu + \nu)(G) > 0$, then we can find compact sets R and S with R contained in G so that $P^x(J(X,R,S)) > 0$ for some x in E. Since $P^x(J(X,R,S))$ is a bounded potential of a continuous additive functional, by (2.1), there are measures μ_{RS} and ν_{RS} on E so that $U\nu_{RS}(x) = V\mu_{RS}(x)$. Moreover, ν_{RS} and μ_{RS} are supported on R. The proof of (1.3) in [12] applies with little change here to show that there is a function f so that for each x in E,

$$(\mu+\nu)\{y: v(x,y) \neq u(x,y)f(y)\} = 0.$$

Moreover,

$$(\mu+\nu)\{f=0\} = 0.$$

Then if r is any positive bounded function on E, $U_A r(x) = Ur\nu(x) = Vrf^{-1}\nu(x)$ and $Vr\mu(x) = Ufr\mu(x)$. So let H(t) be the continuous additive functional of Y(t) with Revuz measure $f^{-1}(x)\ \nu(dx)$, and let G(t) be the continuous additive functional of X(t) with Revuz measure $f(x)\mu(dx)$. Then $U_A r(x) = V_H r(x)$ and $V_B r(x) = U_G r(x)$ on E. If $\Gamma = \Lambda = E$, then A(t), B(t), G(t), and H(t) are all strictly increasing, and (ii) follows easily. Q.E.D.

Note added in proof: Mokobodzki can obtain the main theorem of section 1 using different hypotheses and a different method of proof.

References

1. A. BENVENISTE and J. JACOD. Systèmes de Lévy des processus de Markov. *Invent. Math.* *21* (1973), 183-198.

2. R. M. BLUMENTHAL and R. K. GETOOR. *Markov Processes and Potential Theory.* Academic Press, New York, 1968.

3. R. M. BLUMENTHAL, R. K. GETOOR, and H. P. McKEAN, JR. Markov processes with identical hitting distributions. *Illinois J. Math.* *6* (1962), 402-420.

4. R. M. BLUMENTHAL, R. K. GETOOR, and H. P. McKEAN, JR. A supplement to "Markov processes with identical hitting distributions." *Illinois J. Math.* *7* (1963), 540-542.

5. K. L. CHUNG and R. K. GETOOR. The condenser problem. *Ann. Probab.* *5* (1977), 82-86.

6. K. L. CHUNG and J. B. WALSH. To reverse a Markov process. *Acta Math.* *123* (1970), 225-251.

7. C. CONSTANTINESCU and A. CORNEA. *Potential Theory on Harmonic Spaces.* Springer-Verlag, Berlin, 1972.

8. R. K. GETOOR. *Markov Processes: Ray Processes and Right Processes.* Lecture Notes in Math. *440.* Springer-Verlag, Berlin, 1975.

9. R. K. GETOOR. On the construction of kernels. *Séminaire de Probabilités IX (Univ. Strasbourg)*, pp. 443-463, Lecture Notes in Math. *465.* Springer-Verlag, Berlin, 1975.

10. R. K. GETOOR. Transience and recurrence of Markov processes. *Séminaire de Probabilités XIV*, pp. 397-409. Lecture Notes in Math. *784.* Springer-Verlag, Berlin, 1980.

11. J. GLOVER. Compactifications for dual processes. *Ann. Probab.* *8* (1980), 1119-1134.

12. J. GLOVER. Markov processes with identical hitting probabilities. *Trans. Amer. Math. Soc.* (to appear).

13. J. GLOVER. Markov processes with identical last exit distributions. *Z. Wahrscheinlichkeitstheorie verw. Gebiete* *59* (1982), 67-75.

14. J. GLOVER. Representing last exit potentials as potentials of measures. *Z. Wahrscheinlichkeitstheorie verw. Gebiete* (to appear).

15. P. A. MEYER. Note sur l'interprétation des mesures d'équilibre. *Séminaire de Probabilités V (Univ. Strasbourg)*, pp. 213-236, Lect. Notes in Math. *191.* Springer-Verlag, Berlin, 1971.

16. P. A. MEYER. Le retournement du temps d'après Chung et Walsh.

Séminaire de Probabilités V (Univ. Strasbourg), pp. 213-236. Lect. Notes in Math. *191*. Springer-Verlag, Berlin 1971.

17. D. REVUZ. Mesures associées aux fonctionelles additives de Markov I. *Trans. Amer. Math. Soc. 148* (1970), 501-531.

18. R. SMYTHE and J. B. WALSH. The existence of dual processes. *Inv. Math. 19* (1973), 113-148.

JOSEPH GLOVER
Department of Mathematics
University of Florida
Gainesville, FL 32611

Seminar on Stochastic Processes, 1982
Birkhäuser, Boston, 1983

TOPICS IN ENERGY AND POTENTIAL THEORY*

by

JOSEPH GLOVER

0. Introduction

Energy is a frustratingly delicate item in modern Markov process theory. It is a subject which commands attention, since it is linked so closely with maximum principles and Hunt's hypothesis (H). It entered potential theory in the work of Cartan and Dény, where it enabled them to prove various delicate principles about symmetric potential kernels. It has flourished in the modern theory of Dirichlet spaces and has added to the body of knowledge concerning symmetric Markov processes. But while energy is a natural and cooperative partner in the study of symmetric potential kernels, it becomes increasingly intractable as one attempts to study more asymmetric kernels and processes. Concrete results in this domain are few. We present some topics in energy and potential theory for Markov processes with nonsymmetric potential kernels which complement several results and articles by various authors.

Throughout this paper, we assume that $X = (\Omega, F, F_t, X_t, \theta_t, P^x)$ is a standard Markov process in duality on (E, E) with another standard process $\hat{X} = (\hat{\Omega}, \hat{F}, \hat{F}_t, \hat{X}_t, \hat{\theta}_t, \hat{P}^x)$ with respect to a sigma finite excessive reference measure $m(dx)$ as described in Chapter VI of [2]. Let

*Research supported in part by NSF Grant MCS-8002659.

$u^a(x,y)$ be the regularized a-potential density of the resolvent $U^a(x,dy)$ chosen as in VI-1 in [2]. (We use standard notation from Chapter VI of [2] throughout this paper.) We shall assume that X and $\hat{X}$ are transient so that $u(x,\cdot) < \infty$ a.s. (m) for each x, and $u(\cdot,y) < \infty$ a.s. (m) for each y. We now list several "principles" and discuss their relationships.

(E*a) If μ is a signed measure so that $U^a|\mu|(x)$ is bounded, then $\mu U^a \mu = \int u^a(x,y)\mu(dx)\mu(dy) \geq 0$.

(M*a) If μ is a positive measure so that $U^a\mu(x)$ is bounded, then $\sup\{U^a\mu(x)\colon x \text{ is in } E\} = \sup\{U^a\mu(x)\colon x \text{ is in the Support } (\mu)\}$.

(H) Every semipolar set is polar.

We call (E*), (M*), and (H) the bounded energy principle, the bounded maximum principle, and Hunt's hypothesis (H), respectively. If $u(x,y) = u(y,x)$, then (E*a), (M*a), and (H) are valid for all nonnegative a. The relationship between (M*) and (H) was studied by Blumenthal and Getoor in [1], where they showed that these two principles are equivalent if the excessive functions of X are lower semicontinuous. Their proof that (H) implies (M*) depends on the fact that if $U\mu$ is bounded, then μ does not charge polar sets —provided the excessive functions are lower semicontinuous. In section 1, we give a very simple example where $U\mu$ is bounded, μ charges a polar set, and the excessive functions are not lower semicontinuous. This leaves the following question unanswered: are (M*) and (H) equivalent if X and $\hat{X}$ are standard processes?

Say that the fine and cofine topologies differ by semipolar (resp. polar) sets if the fine and cofine interiors of an arbitrary subset of E differ by a semipolar (resp. polar) set (or equivalently, the fine and cofine closures of an arbitrary subset of E differ by a semipolar (resp. polar) set). Blumenthal and Getoor [1] showed that if excessive functions are lower semicontinuous, then (H), and hence (M*), are equivalent to the fine and cofine topologies differing by polar sets. If we drop the lower semicontinuity hypothesis, we can prove that (H) is equivalent to the fact that the fine and cofine topologies differ by polar sets (see section 2). Finally, Rao [7] showed that (E*) implies (M*).

These equivalences are of interest to the probabilist because they provide different methods of deciding when Hunt's hypothesis (H) is valid. There are few concrete results for nonsymmetric kernels and processes. Rao [7] gave a simple proof of the Kanda [6] - Forst [3] theorem by showing that (E*) and hence (H) hold for Lévy processes in R^d with Lévy-Khinchin exponents $\Psi(x)$ satisfying $|1 + \mathrm{Re}\,\psi(x)| \geq M|\mathrm{Im}\,\psi(x)|$ for some positive constant M. Thus the difficult question of which nonsymmetric Lévy processes satisfy (E*) and (H) remains largely unanswered.

In the classical work on energy and potential theory, one does not find the bounded energy and maximum principles, but the full energy and maximum principles:

(E) If μ is a signed measure so that $|\mu|U|\mu| < \infty$, then $\mu U\mu \geq 0$.

(M) If μ is a positive measure, then

$$\sup\{U\mu(x) : x \text{ is in } E\} = \sup\{U\mu(x) : x \text{ is in Support } (\mu)\}.$$

In [5], we gave a probabilistic proof that (E) implies (M), assuming that the processes X and $\hat{X}$ have infinite lifetimes, an invariant duality measure m(dx), and lower semicontinuous excessive functions. In section 3, we give a potential-theoretic proof that (E) implies (M), assuming only that X and $\hat{X}$ are transient standard processes in duality with lower semicontinuous excessive functions. We verified in [5] that (E) is true for a very small class of nonsymmetric Lévy processes ((E) is true, of course, for all symmetric Lévy processes).

1. Bounded Potentials of Measures: An Example

Let $E = \{(x,y): x^2 + y^2 = 1\}$. Let $X(t)$ be the process on E which is uniform motion with speed 1 counterclockwise around E until it reaches the point (1,0), when it dies and goes to a cemetery Δ. Note that $P^{(1,0)}(\zeta = 2\pi) = 1$ and $X(\zeta-) = (1,0)$ a.s. (P^x) for all x in E. Let $\hat{X}(t)$ be the process on E which is uniform motion with speed 1 clockwise around E until it reaches the point (1,0) when it dies. Then $X(t)$ and $\hat{X}(t)$ are standard processes in duality with respect to Lebesgue measure on E. It is simple to check that the point (1,0) is polar, and $u((x,y),(1,0)) = \int u((x,y),(u,v))\varepsilon_{(1,0)}(du,dv)$ is bounded.

2. The Fine and Cofine Topologies

The proof of (2.2) is a modification of the proof of (5.12) in [1]. Recall the following result from [1].

(2.1) PROPOSITION. *The fine and cofine topologies always differ by semipolar sets.*

(2.2) THEOREM. (H) *is equivalent to the fact that the fine and cofine topologies differ by polar sets.*

PROOF. If (H) holds, then it is clear that the fine and cofine topologies differ by polar sets by (2.1). Now assume that the fine and cofine topologies differ by polar sets, and let K be a compact totally thin set so that $H(x) = E^x(e^{-T(K)}) < b < 1$ on K. Now $H(x) = U^1\pi(x)$, where $\pi(dx)$ is the 1-capacitary measure of K and does not charge polar sets. Let $e > 0$, and let $B = \{H(x) < b + e\}$. Then B is finely open and contains K. If D is the cofine interior of B, then $B - D$ is polar,

$\pi(B-D) = 0$, and

$$\hat{P}^1_B\pi(G) = \int_{D\cap K} \hat{P}^1_B(G,x)\,\pi(dx) = \pi(G)$$

for all G in E. Therefore, $P^1_B U^1\pi = U^1\hat{P}^1_B\pi = U^1\pi$. Thus $H(x) \leq b+e$ for all $e > 0$, so $H(x) \leq b$ everywhere. If K is not polar, there is an x in K^c so that $P^x(T(K) < \infty) > 0$. By Shih's theorem [4], we may choose a sequence $(G(n))$ of finely open sets containing K so that $T(n) \equiv T(G(n))$ increases to $T(K)$ a.s. (P^x). We claim a.s. (P^x) on $\{T(K)<\infty\}$, $T(n)<T(K)$ for all n. To see this, we modify the proof of (5.4) in [1].

(2.3) LEMMA. $T(K)\circ\theta_{T(n)} = 0$ a.s. *on* $\{T(n) = T(K) < \infty\}$.

PROOF. The statement is equivalent to $P^1_{G(n)}P^1_K 1 = P^1_K 1$. Take a sequence (h_k) of bounded positive measurable functions so that $U^1 h_k$ increases to 1 and let $dm(k) = h_k dm$. Then

$$P^1_{G(n)}P^1_K 1 = \lim_{k\to\infty} P^1_{G(n)}P^1_K U^1 h_k = \lim_{k\to\infty} U^1\hat{P}^1_{G(n)}\hat{P}^1_K m(k).$$

If $H(n)$ denotes the cofine interior of $G(n)$, then $G(n)-H(n)$ is polar. Thus,

$$\lim_{k\to\infty} U^1\hat{P}^1_{G(n)}\hat{P}^1_K m(k) = \lim_{k\to\infty} U^1\hat{P}^1_{H(n)}\hat{P}^1_K m(k).$$

But $\hat{P}^1_K m(k)$ is carried by $K \cup {}^rK$, and $(K\cup {}^rK)\cap H(n)^c$ is polar. Therefore,

$$U^1\hat{P}^1_{H(n)}\hat{P}^1_K m(k) = U^1\hat{P}^1_K m(k).$$

So $P^1_{G(n)}P^1_K 1 = \lim_{k\to\infty} U^1\hat{P}^1_K m(k) = \lim_{k\to\infty} P^1_K U^1 h_k = P^1_K 1.$ Q.E.D.

Now we may complete the proof of (2.2). Since K is thin, $T(K)\circ\theta_{T(n)} > 0$ a.s. (P^x) on $\{T(n) = T(K) < \infty\}$, and therefore, $P^x(T(n) = T(K) < \infty) = 0$. By (5.9i) in [1], $H(X_{T(n)})$ converges to 1 a.s. (P^x) on $\{T(K) < \infty\}$. But $H(x) \leq b < 1$. Thus K must be polar, and thus (H) must hold. Q.E.D.

3. The Maximum Principle

In this section, we assume that X and $\hat{X}$ are transient standard processes in duality, the excessive functions are lower semicontinuous, and (E) holds. Rao [7] showed that (E*) implies (M*), and Blumenthal and Getoor [1] showed that (M*) implies (H) under these hypotheses. Blumenthal and Getoor also showed that (M) is equivalent to the following principle.

(P) If μ is a finite measure with compact support so that $\mu U\mu < \infty$, then μ does not charge semipolar sets.

Thus, in order to show that (E) implies (M), it suffices to show that (E) implies (P).

Let $(n,v) = nUv = \int u(x,y)n(dx)v(dy)$, and let $\|n\| = (n,n)^{\frac{1}{2}}$.

(3.1) LEMMA. *If* n *and* v *are signed measures with* $(|v|,|v|) < \infty$ *and* $(|n|,|n|) < \infty$, *then* $|(n,v) + (v,n)| \leq 2\,\|n\|\,\|v\|$.

PROOF. Set $b = -((n,v) + (v,n))/2(n,n)$. Then,

$$0 \leq (bn + v, bn + v) = 4(n,n)(v,v) - ((n,v) + (v,n))^2. \qquad \text{Q.E.D.}$$

(3.2) THEOREM. (E) *implies* (M).

PROOF. Let μ be a positive finite measure supported on a compact polar set K, and suppose $(\mu,\mu) < \infty$. Let $(G(n))$ be a sequence of finely open sets decreasing to K so that $\overline{G(n)}$ is compact for all n. Each G(n) may be chosen so that $\sup\{t: X(t) \in G(n)\} < \infty$ a.s. Let $\pi(n)$ be the capacitary measure of G(n): $P_{G(n)}1 = U\pi(n)$. Then $\pi_n(1)$ decreases to zero. Since (H) holds, $\pi_n(1) = (\pi_n,\pi_n)$. Choose a subsequence of (π_n), say (γ_n), so that if $\gamma = \sum \gamma_n$, then $\gamma U\gamma < \infty$. (To do this, proceed inductively as follows. Let $\gamma_1 = \pi_1$, and let $A = \gamma_1(1)$. Then $(\gamma_1,\gamma_1) < 2A$. Assume $(\gamma_i)_{i \leq n}$ has been chosen so that $\|\sum_{i\leq n} \gamma_i\|^2 < 2A$. Then $\|\pi_k + \sum_{i\leq n} \gamma_i\|^2 < 2\|\pi_k\| \, \|\sum_{i\leq n} \gamma_i\| + \|\sum_{i\leq n} \gamma_i\|^2 + \|\pi_k\|^2$. Since $\pi_k(1)$ decreases to zero, one can choose k sufficiently large so that the right hand side of the last inequality is less than 2A. Set $\gamma_{n+1} = \pi_k$.) Since $U\pi_n = 1$ on K, $U\gamma = \infty$ on K. If K is not empty, then $\infty = \mu U\gamma \leq 2\|\mu\| \, \|\gamma\| < \infty$. Thus K must be empty and $\mu = 0$. This proves (P) which implies (M). Q.E.D.

In the classical literature, (M) is called the first maximum principle. We showed in [5] that (P) also implies the second maximum principle.

(N) Let μ be a finite measure with compact support K and $\mu U\mu < \infty$. If $U\mu \leq U\nu$ on K, then $U\mu \leq U\nu$ everywhere.

(3.3) COROLLARY. (E) *implies* (N).

References

1. R. M. BLUMENTHAL and R. K. GETOOR. Dual processes and potential theory. Proc. 12th Biennial Seminar of the Canadian Math. Congress, Canadian Math. Soc. (1970), 137-156.

2. R. M. BLUMENTHAL and R. K. GETOOR. *Markov Processes and Potential Theory*. Academic Press, New York, 1968.

3. G. FORST. The definition of energy in non-symmetric translation invariant Dirichlet spaces. *Math. Ann. 216* (1975), 165-172.

4. R. K. GETOOR. *Markov Processes: Ray Processes and Right Processes.* Lecture Notes in Math. *440*. Springer-Verlag, Berlin, 1975.

5. J. GLOVER. Energy and the maximum principle for nonsymmetric Hunt processes. *Teor. Verojatnost. ii Primenen. 26* (1981) no. 4, 757-768.

6. M. KANDA. Two theorems on capacity for Markov processes with independent increments. *Z. Wahrscheinlichkeitstheorie verw. Gebiete*, *35* (1976), 159-166.

7. K. M. RAO. On a result of M. Kanda. *Z. Wahrscheinlichkeitstheorie verw. Gebiete*, *41* (1977), 35-37.

JOSEPH GLOVER
Department of Mathematics
University of Florida
Gainesville, FL 32611

Seminar on Stochastic Processes, 1982
Birkhäuser, Boston, 1983

ON THE p-VARIATION OF GAUSSIAN RANDOM FIELDS WITH SEPARABLE INCREMENTS

by

DITLEV MONRAD*

1. Introduction

The p-variation of the sample functions of a Gaussian random process $\{X(t,\omega): 0 \leq t \leq 1\}$ has been studied in a number of papers. (See [5], [6] and [7].) In this paper we study the p-variation of the sample functions of Gaussian random fields.

Let $I_d = [0,1]^d$ be the unit cube in R^d. A point $t \in I_d$ is written explicitly as $t = (t_1,t_2,\ldots,t_d)$. The points $(0,0,\ldots,0)$ and $(1,1,\ldots,1)$ in I_d will be written 0 and 1, respectively. To each $a,b \in I_d$ with $a_i < b_i$ for $i = 1,\ldots,d$ there corresponds the *half-open interval*

$$(1.1)\quad (a,b] = \{t \in I_d: a_i < t_i \leq b_i \quad \text{for } i = 1,\ldots,d\}.$$

For $A = (a,b]$, write $|A| = \Pi(b_i - a_i)$. And for any real-valued function f on I_d, we define f(A), the increment of f over A, as follows

$$(1.2)\quad f(A) = F_0 - F_1 + \cdots + (-1)^d F_d,$$

*Supported in part by the National Science Foundation.

where F_k is the sum of all $\binom{d}{k}$ terms of the form $f(c_1, \ldots, c_d)$ with $c_i = a_i$ for exactly k indices $i \in \{1,2,\ldots,d\}$ and $c_i = b_i$ for the remaining $d-k$ indices. In particular, if $d = 2$, then

$$(1.3) \quad f(A) = f(b_1, b_2) - f(a_1, b_2) - f(b_1, a_2) + f(a_1, a_2).$$

For $p \geq 1$ and any finite partition π: $A_1, \ldots, A_m$ of $I_d^+ = \{t \in I_d: t_i > 0, i = 1,\ldots,d\}$ into half-open rectangles, put

$$(1.4) \quad V_p(f;\pi) = \sum_{k=1}^{m} |f(A_k)|^p.$$

Generalizing Vitali's notion of variation we define for $p \geq 1$ the p-variation of f,

$$(1.5) \quad V_p(f) = \sup_{\pi} V_p(f;\pi)$$

where the supremum is taken over all finite partitions π.

Now consider a separable, *centered* Gaussian random field $X = \{X(t,\omega): t \in I_d\}$ on a complete probability space (Ω, F, P). We wish to study the p-variation of the sample functions $t \to X(t,\omega)$. We shall say that X has *separable increments* if there exists a countable, dense subset $S \subset I_d$ such that for a.a. ω, any finite partition π: $(a^1, b^1], \ldots, (a^m, b^m]$ of I_d^+, and any $\varepsilon > 0$, there exists a partition π': $(c^1, d^1], \ldots, (c^m, d^m]$ such that for $k = 1, \ldots, m$, the points c^k and d^k are in S, $|c^k - a^k| < \varepsilon$, $|d^k - b^k| < \varepsilon$, and $|X((a^k, b^k], \omega) - X((c^k, d^k], \omega)| < \varepsilon$.

If X has separable increments, then the p-variation $V_p(X(\cdot,\omega))$ is a random variable with values in $[0,\infty]$.

Separability alone is not enough to ensure that X has separable increments. In fact, the p-variation of a separable random field X need not be measurable, and different separable modifications of X may have different p-variations. (Following the standard terminology, we call two random fields Z_1 and Z_2 *modifications* of each other if for each t, $P\{Z_1(t) = Z_2(t)\} = 1$.)

Define, for every half-open rectangle $A \subset I_d$,

(1.6) $$\sigma(A) = E|X(A)|.$$

We shall use the notation

(1.7) $$\log^*(s) = \max\{1, |\log(s)|\}, \quad s > 0$$

(1.8) $$\log_2^*(s) = \log^*(\log^*(s)), \quad s > 0.$$

For $p \geq 1$ define

(1.9) $$G(p) = \sup_\pi \sum_k \sigma(A_k)^p$$

(1.10) $$G_1(p) = \sup_\pi \sum_k \sigma(A_k)^p(\log^*(\sigma(A_k)))^{p/2}$$

(1.11) $$G_2(p) = \sup_\pi \sum_k \sigma(A_k)^p(\log_2^*(\sigma(A_k)))^{p/2}$$

where the supremum is taken over all finite partitions $\pi: A_1, \ldots, A_m$ of I_d^+ into half-open rectangles (and the summand is 0 if $\sigma(A_k) = 0$). Clearly

(1.12) $$G(p) \leq G_2(p) \leq G_1(p).$$

If $G(p) < \infty$ for some p, then $G_1(p') < \infty$ for all $p' > p$. Define

(1.13) $$\gamma = \inf\{p \geq 1: G(p) < \infty\}$$

(with the understanding that $\gamma = \infty$ if the set is empty).

THEOREM 1. *If* $\gamma < \infty$, *then* X *has a separable modification with separable increments.*

The p-variation of a random field X with separable increments is obviously minimal in the sense that the p-variation of any modification of X is at least as great with probability 1.

THEOREM 2. *Let* X *be a centered Gaussian random field with separable increments. Then* $V_p(X) < \infty$ *with either probability one or with probability zero. If* $V_p(X) < \infty$ *a.s., then there exists a constant* $\varepsilon > 0$ *such that* $E[\exp(\varepsilon V_p(X)^{2/p})] < \infty$.

THEOREM 3. *Let* X *be a centered Gaussian random field. If* $G(p) = \infty$, *then* $V_p(X) = \infty$ *a.s.*

So $G(p) < \infty$ is a necessary condition for X to have sample functions of finite p-variation.

THEOREM 4. *For a centered Gaussian random field* X *with separable increments we have* $V_1(X) < \infty$ *a.s. if and only if* $G(1) < \infty$.

THEOREM 5. *Let* $p > 1$. *If* X *has separable increments and* $G_1(p) < \infty$, *then* $V_p(X) < \infty$ *a.s.*

Combining Theorems 3 and 5 we get

COROLLARY 1. *If* X *has separable increments then*

(1.14) $\gamma = \inf\{p \geq 1 : V_p(X) < \infty \text{ a.s.}\}$.

The number γ is called the *variation dimension* of X.

COROLLARY 2. *If* X *is separable and for some* $p > d$,

$$E|X(s) - X(t)| = O(|s - t|^{d/p}|\log|s - t||^{-\frac{1}{2}})$$

as $|s - t| \downarrow 0$, *then* $V_p(X) < \infty$ *a.s.*

REASON: Consider any half-open rectangle $A \subset I_d$. Let the shortest edge of A have length ℓ. Then $\sigma(A) \leq C\ell^{d/p}|\log(\ell)|^{-\frac{1}{2}}$ and $\ell^d \leq |A|$. It follows that

$$\sigma(A)^p(\log^*(\sigma(A)))^{p/2} \leq C'|A|.$$

This shows that $G_1(p) < \infty$.

CONJECTURE. If X is centered, has separable increments, and $G_2(p) < \infty$, then $V_p(X) < \infty$ a.s.

We shall only prove this under the additional assumption that X has *stationary increments,* in the sense that

$$E[X((a,b])^2] = E[X((a + h, b + h])^2]$$

whenever $(a,b] \subset I_d$ and $(a + h, b + h] \subset I_d$.

The paper is organized as follows: Theorem 1 is proved in Section 2. Theorems 2, 3 and 4 are established in the same way as in the one-parameter case. (See [4] and [5].) We get around the lack of measurability in Theorem 3 by restricting the random field to a suitable, countable subset of I_d. We omit the details. In Section 3 we prove Theorem 5 and show that for random fields with stationary increments, $G_2(p) < \infty$ is a sufficient condition for ensuring sample functions of finite p-variation. Section 4 contains some examples.

Throughout, the letters c and c_i will denote positive constants. Their values are unimportant and may change from one context to another, even from line to line.

2. Existence of modifications with separable increments

We shall prove Theorem 1. Assume that $G(p) < \infty$ for some $p \geq 1$. Define for $i = 1,\dots,d$ and $0 < u \leq 1$,

$$(2.1) \qquad F_i(u) = \sup_{\pi} \sum_k E[|X(A_k)|^p],$$

where the sup is taken over all finite partitions $\pi: A_1,\dots,A_m$ of $\{t \in I_d^+ : 0 < t_i \leq u\}$ into half-open rectangles.

The functions F_i are increasing and bounded. And for any two points $s = (s_1,\dots,s_d)$ and $t = (t_1,\dots,t_d)$ in I_d^+ we have

$$(2.2) \quad E[|X((o,s]) - X((o,t])|^p] \leq d^{p-1} \sum_{i=1}^{d} (F_i(s_i \vee t_i) - F_i(s_i \wedge t_i)),$$

where for any real numbers α and β we write $\alpha \vee \beta = \max\{\alpha,\beta\}$ and $\alpha \wedge \beta = \min\{\alpha,\beta\}$.

For $i = 1,\dots,d$, let D_i denote the countable set of points of discontinuity for F_i. Let Q denote the rational numbers in $[0,1]$. Put

$$(2.3) \qquad S = (D_1 \cup Q) \times \cdots \times (D_d \cup Q) \subset I_d.$$

It follows from (2.2) that each variable $X(A)$ can be approximated in L^2 by variables $X(A_n)$ where the rectangles A_n have all their corners in S. The L^2-norm defines a metric on the set $\{X(A): A \subset I_d\}$ parametrized by the collection of all half-open rectangles A contained in I_d.

For every $\varepsilon > 0$, let $N(\varepsilon)$ denote the minimal number of balls of L^2-radius $\leq \varepsilon$ needed to cover the set of random variables $\{X(A): A \subset I_d\}$. From (2.2) it follows that

$$(2.4) \qquad N(\varepsilon) \leq C\varepsilon^{-dp}$$

for some constant $C > 0$. It therefore follows from Dudley's entropy-theorem ([2]) that $\{X(A): A \subset I_d\}$ has a modification $\{Y(A,\omega): A \subset I_d\}$ such that for a.a. ω, there exists a $\delta(\omega) > 0$ such that for any rectangles A and B satisfying

$$E|X(A) - X(B)| \leq \delta \leq \delta(\omega) \tag{2.5}$$

we have

$$|Y(A,\omega) - Y(B,\omega)| \leq C \int_0^\delta (\log N(\varepsilon))^{\frac{1}{2}} d\varepsilon. \tag{2.6}$$

If we write $d(A,B) = E|X(A) - X(B)|$ and use the estimate (2.4), we have

$$|Y(A,\omega) - Y(B,\omega)| \leq C' d(A,B)|\log^*(d(A,B))|^{\frac{1}{2}} \tag{2.7}$$

for all A and B with $d(A,B) \leq \delta(\omega)$. Now define the random field $\{\tilde{Y}(t): t \in I_d\}$ by

$$\tilde{Y}(t) = \begin{cases} 0 & \text{if any } t_i = 0 \\ Y((o,t]) & \text{if } \quad t \in I_d^+ \end{cases} . \tag{2.8}$$

The increment of the random field $\tilde{Y}$ over a given rectangle $A \subset I_d$ is a.s. $Y(A)$. More precisely, with probability one,

$$\tilde{Y}(A) = Y(A) \tag{2.9}$$

for every rectangle $A \subset I_d$. It is clear that (2.9) holds a.s. for all the countably many rectangles A with corners in S. It then follows from (2.2) and (2.7) that (2.9) must hold simultaneously for all rectangles A.

It now follows from (2.2) and (2.7) that the random field $\tilde{Y}$ has separable increments.

We complete the proof of Theorem 1 as follows: For $t \in I_d - I_d^+$ define $\tilde{X}(t,\omega)$ such that $\{\tilde{X}(t): t \in I_d - I_d^+\}$ is a separable modification of $\{X(t): t \in I_d - I_d^+\}$. For $k = 1,\ldots,d$ and $t \in I_d^+$ let $\tilde{X}_k(t)$ be the sum of all $\binom{d}{k}$ terms of the form $\tilde{X}(s_1,\ldots,s_d)$, where $s_i = 0$ for exactly k indices in $\{1,\ldots,d\}$ and $s_i = t_i$ for the remaining $d-k$ indices. Now define for $t \in I_d^+$,

$$\tilde{X}(t) = \tilde{Y}(t) + \sum_{k=1}^{d} (-1)^{k-1} \tilde{X}_k(t).$$

3. Conditions for $V_p(X) < \infty$ a.s.

It follows from (2.2) and (2.7) that in order to prove Theorem 5 we only have to show that $G_1(p) < \infty$ implies that $V_p(\tilde{Y}) < \infty$ a.s., where $\tilde{Y}$ is defined by (2.8).

Consider the $3^d - 1$ d-tuples $\underset{\sim}{R} = (R_1,\ldots,R_d)$, where each R_i is one of the three relations $<$, $=$, or $>$, and not all R_i are $=$. Each such d-tuple $\underset{\sim}{R}$ generates a partial ordering on I_d,

$$s\underset{\sim}{R}t \iff s_i R_i t_i \quad \text{for } i = 1,\ldots,d. \tag{3.1}$$

Combining the estimates (2.2) and (2.7) we see that for a.a. ω, the limit

$$\lim_{u \to t,\, t\underset{\sim}{R}u} \tilde{Y}(u,\omega) \tag{3.2}$$

exists for each of the $3^d - 1$ order relations $\underset{\sim}{R}$ at every point $t \in I_d$ for which the set $\{u \in I_d: t\underset{\sim}{R}u\}$ is nonempty.

It follows that the sample functions of $\tilde{Y}$ are a.s. bounded.

Combining (2.7) with the fact that there exist rectangles B for which $\sigma(B)$ is as small as you like, we see that for any rectangle A with $\sigma(A) < \delta(\omega)$ we have

(3.3) $$|\tilde{Y}(A,\omega)| \leq C'\sigma(A)|\log^*(\sigma(A))|^{\frac{1}{2}}.$$

Combining this inequality with the fact that $\tilde{Y}$ has bounded sample functions, we see that for a.a. ω there exists $C'(\omega)$ such that for all rectangles $A \subset I_d$,

(3.4) $$|\tilde{Y}(A,\omega)| \leq C'(\omega)\sigma(A)|\log^*(\sigma(A))|^{\frac{1}{2}}.$$

It immediately follows from (3.4) that if $G_1(p) < \infty$, then $V_p(\tilde{Y}) < \infty$ a.s. This completes the proof of Theorem 5.

We shall now take a closer look at the variation of Gaussian random fields.

LEMMA 1. Let ψ be a nondecreasing, continuous function on $[0,\infty)$ with $\psi(0) = 0$ such that

(3.5) $$\int_0^\infty \psi(he^{-x^2})dx = O(\psi(h)), \quad \text{as } h \downarrow 0$$

(3.6) $$u\psi(v) \leq c_1\psi(uv), \quad \text{for } 0 \leq u \leq 1 \text{ and } v \geq 0.$$

Let $\{X(t)\colon t \in I_d\}$ be a separable centered Gaussian random field. Assume that

(3.7) $$E|X(s) - X(t)| \leq c_2\psi\Big(\sum_{i=1}^{d} (F_i(s_i \vee t_i) - F_i(s_i \wedge t_i))\Big)$$

for some constant $c_2 > 0$ and some nondecreasing functions F_i on $[0,1]$ with $F_i(0) = 0$. For $h > 0$ put

(3.8) $$I(h) = \{t \in I_d\colon F_i(t_i) \leq h \quad \text{for } i = 1,\ldots,d\}.$$

There exists a constant $c_3 > 0$ (depending only on ψ and c_2) such that for $h > 0$ and $x \geq 1$,

$$(3.9) \qquad P\{\sup_{s,t \in I(h)} |X(s) - X(t)| > c_3\psi(h)x\} \leq c_3 \exp(-x^2/2).$$

PROOF. By arguments similar to those in the proof of Lemma 2.4 in [5] we can show that there exists a centered, stochastically continuous Gaussian random field $\{Y(t): t \in R^d\}$ satisfying

$$(3.10) \qquad E|Y(s) - Y(t)| \leq c_4\psi(|s - t|),$$

such that with probability one,

$$(3.11) \qquad X(t,\omega) = Y(F(t),\omega)$$

for all $t \in I_d$, where $F(t) = (F_1(t_1),\dots,F_d(t_d))$. We then complete the proof by applying Fernique's lemma (see for example Lemma 1.1 on page 138 in [3]) to the process

$$Z(s,t) = Y(s) - Y(t), \quad s,t \in [0,h]^d.$$

LEMMA 2. Let ϕ be a strictly increasing continuous function on $[0,\infty)$ with $\phi(0) = 0$. Assume that ψ, the inverse of ϕ, satisfies (3.5) and (3.6). Let $\{X(t): t \in I_d\}$ be a centered Gaussian random field with separable increments. For any half-open rectangle $A \subset I_d$ put

$$(3.12) \qquad \Phi_A = \Phi(A) = \sup_{\pi} \sum_k \phi(\sigma_X(A_k)),$$

where the supremum is taken over all finite partitions π: $A_1,\dots,A_m$ of A into half-open rectangles.

Then there exists a constant c_5 (depending only on ϕ) such that for $x \geq 1$ and any rectangle $A \subset I_d$ with $\Phi_A < \infty$,

$$(3.13) \qquad P\{\sup_{A' \subset A} |X(A')| > c_5\psi(\Phi_A)x\} \leq \exp(-x^2/2).$$

PROOF. It is obvious that we may assume that $A = I_d^+$. For $i = 1,\ldots,d$ and $0 < u \leq 1$ define

$$(3.14) \qquad F_i(u) = \sup_{\pi} \sum_k \phi(\sigma_X(A_k)),$$

where the sup is taken over all finite partitions π of $\{t \in I_d^+ : t_i \leq u\}$. Then

$$(3.15) \qquad E|X(s) - X(t)| \leq \Sigma\psi(|F_i(s_i) - F_i(t_i)|)$$
$$\leq d\psi(\Sigma|F_i(s_i) - F_i(t_i)|).$$

We can therefore apply the previous lemma with $h = \Phi_A$. For $x \geq 1$,

$$P\{\sup_{A' \subset A} |X(A')| > 2^{d-1}c_3\psi(h)x\}$$
$$\leq P\{\sup_{s,t \in I(h)} |X(s) - X(t)| > c_3\psi(h)x\} \leq c_3\exp(-x^2/2).$$

We can now prove

THEOREM 6. *Let* X *be a centered Gaussian random field on* I_d *with separable increments. If* $G_2(p) < \infty$ *and* X *has stationary increments, then* $V_p(X) < \infty$ *a.s.*

PROOF. For any integer valued vectors $n = (n_1,\ldots,n_d)$ and $j = (j_1,\ldots,j_d)$, where $n_i = 1,2,\ldots$ and $j_i = 0,1,\ldots,2^{n_i+1} - 2$, put

$$J_{n,j} = \{t \in I_d : j_i\, 2^{-n_i-1} < t_i \le (j_i+2)2^{-n_i-1},\quad i=1,\dots,d\}. \tag{3.16}$$

Let $\phi(u) = u^p(\log_2^*(u))^{p/2}$, for $u > 0$. For any rectangle $A \subset I_d$, define $\Phi(A)$ by (3.12). Since X has stationary increments,

$$\Phi(J_{n,j}) \le \Phi(I_d^+)2^{-(n_1+\cdots+n_d)}. \tag{3.17}$$

If ψ denotes the inverse of ϕ, then

$$\psi(\Phi(J_{n,j})) \le c_6\, 2^{-(n_1+\cdots+n_d)/p}(\log(n_1+\cdots+n_d))^{-\frac{1}{2}}. \tag{3.18}$$

Put $c_7 = c_5c_6$, where c_5 is the constant in (3.13). Let m and k be positive integers such that

$$p/2 + 2d < m + d < k. \tag{3.19}$$

Put

$$\Lambda_{n,j} = \{\omega : \sup_{A \subset J_{n,j}} |X(A,\omega)| > c_7\sqrt{2k}\, 2^{-(n_1+\cdots+n_d)/p}\}. \tag{3.20}$$

It follows from Lemma 2 that

$$P(\Lambda_{n,j}) \le (n_1+\cdots+n_d)^{-k}. \tag{3.21}$$

Define

$$Z_n(\omega) = \#\{j : \omega \in \Lambda_{n,j}\}. \tag{3.22}$$

It follows from (3.16) and (3.21) that

$$E[Z_n] \le 2^{n_1+\cdots+n_d+d}(n_1+\cdots+n_d)^{-k}. \tag{3.23}$$

By Markov's inequality,

$$(3.24)\qquad \sum_{n_1=1}^{\infty}\cdots\sum_{n_d=1}^{\infty} P\{Z_n > 2^{n_1+\cdots+n_d}(n_1+\cdots+n_d)^{-m}\}$$

$$\leq \sum_{n_1}\cdots\sum_{n_d} 2^d(n_1+\cdots+n_d)^{-(k-m)} < \infty.$$

By Borel-Cantelli, there exists for a.a. ω a finite number $C(\omega)$ such that for all n,

$$(3.25)\qquad Z_n(\omega) \leq C(\omega)2^{n_1+\cdots+n_d}(n_1+\cdots+n_d)^{-m}.$$

Now consider a finite partition π: $(a^1,b^1],\ldots,(a^N,b^N]$ of I_d^+. To simplify notation, write $A_\ell = (a^\ell,b^\ell]$, for $\ell = 1,\ldots,N$. Put

$$(3.26)\qquad \Lambda(\omega) = \{\ell: |X(A_\ell)|^p > 4^d(c_7\sqrt{2k})^p|A_\ell|\}.$$

Obviously,

$$(3.27)\qquad \sum_{\ell\notin\Lambda(\omega)} |X(A_\ell,\omega)|^p \leq 4^d(c_7\sqrt{2k})^p.$$

For $n = (n_1,\ldots,n_d)$, put

$$(3.28)\qquad \Lambda_n = \{\ell: 2^{-n_i-2} \leq b_i^\ell - a_i^\ell < 2^{-n_i-1},\quad i = 1,\ldots,d\}.$$

If $\ell \in \Lambda(\omega) \cap \Lambda_n$, then $A_\ell \subset J_{n,j}$ and $\omega \in \Lambda_{n,j}$, for some j. For fixed j

$$(3.29)\qquad \#\{\ell \in \Lambda_n: A_\ell \subset J_{n,j}\} \leq 4^d.$$

And by definition,

$$(3.30)\qquad \#\{j: \omega \in \Lambda_{n,j}\} = Z_n(\omega).$$

It follows that

(3.31) $\#(\Lambda(\omega) \cap \Lambda_n) \leq 4^d Z_n(\omega).$

Using the estimates (3.25) and (3.4) we get

$$\sum_{\ell \in \Lambda(\omega)} |X(A_\ell,\omega)|^p = \sum_{n_1} \cdots \sum_{n_d} \sum_{\ell \in \Lambda(\omega) \cap \Lambda_n} |X(A_\ell,\omega)|^p$$

$$\leq \sum_{n_1} \cdots \sum_{n_d} \sum_{\ell \in \Lambda(\omega) \cap \Lambda_n} [C'(\omega)\sigma(A_\ell)(\log^*(\sigma(A_\ell)))^{\frac{1}{2}}]^p$$

$$\leq \sum_{n_1} \cdots \sum_{n_d} C(\omega) Z_n(\omega) 2^{-(n_1 + \cdots + n_d)} (n_1 + \cdots + n_d)^{p/2}$$

$$\leq \sum_{n_1} \cdots \sum_{n_1} C''(\omega)(n_1 + \cdots + n_d)^{-(m - p/2)} < \infty.$$

Combining this with (3.27) we see that $V_p(X) < \infty$ a.s. This completes the proof of Theorem 6.

4. Examples

Consider first a separable, centered Gaussian process $\{X(t,\omega): t \in I_2\}$ with covariance

$$E[X(s_1,s_2)X(t_1,t_2)] = \exp\{-((s_1 - t_1)^2 + (s_2 - t_2)^2)^{\alpha/2}\},$$

for some α, $0 < \alpha \leq 2$. We shall show that if $\alpha < 2$, then the variation dimension γ equals $4/\alpha$.

For small $h > 0$, let $a = (a_1,a_2) \in I_2$ and $b = (a_1 + h, a_2 + h) \in I_2$. Then

$$\sigma^2((a,b]) = 4 - 8\exp(-h^\alpha) + 4\exp(-2^{\alpha/2}h^\alpha) \approx 4(2 - 2^{\alpha/2})h^\alpha.$$

Combining this estimate with Corollary 2, we see that $\gamma = 4/\alpha$. (See also [9].)

If $\alpha = 2$, then $\gamma = 1$, as the next example shows:

Let $\{X(t,\omega): t \in I_2\}$ be a separable, centered Gaussian process with covariance

$$E[X(s_1,s_2)X(t_1,t_2)] = \exp\{-|s_1 - t_1|^\alpha - |s_2 - t_2|^\beta\}$$

where $0 < \alpha \leq \beta \leq 2$. For any rectangle $(a,b]$

$$\sigma^2((a,b]) = 4(1 - \exp\{-|b_1 - a_1|^\alpha\})(1 - \exp\{-|b_2 - a_2|^\beta\})$$

$$\approx 4|b_1 - a_1|^\alpha |b_2 - a_2|^\beta.$$

This shows that $\gamma = 2/\alpha$.

References

[1] S. M. BERMAN. A version of the Lévy-Baxter theorem for the increments of Brownian motion of several parameters. *Proc. Amer. Math. Soc. 18* (1967), 1051-1055.

[2] R. M. DUDLEY. Sample functions of the Gaussian process. *Ann. Probab. 1* (1973), 66-103.

[3] N. C. JAIN and M. B. MARCUS. Continuity of subgaussian processes. *Adv. Probab. 4* (1978), 81-196.

[4] N. C. JAIN and D. MONRAD. Gaussian quasimartingales. *Z. Wahrscheinlichkeitstheorie verw. Gebiete, 59* (1982), 139-159.

[5] N. C. JAIN and D. MONRAD. Gaussian measures in B_p. *Ann. Probab.* To appear.

[6] T. KAWADA and N. KONO. On the variation of Gaussian processes. Proc. Second Japan - USSR Sympos. (Kyoto), pp. 176-192. Lecture

Notes in Math. *330*, Springer-Verlag, Berlin, 1973.

[7] S. J. TAYLOR. Exact asymptotic estimates of Brownian path variation. *Duke Math. J. 39* (1972), 219-241.

[8] L. YODER. Variation of multiparameter Brownian motion. *Proc. Amer. Math. Soc. 46* (1974), 302-309.

[9] N. M. ZINČENKO. On the p-variation of Gaussian random fields. *Theo. Prob. Math. Stat. 19* (1980), 81-86.

DITLEV MONRAD
Department of Mathematics
University of Illinois, Urbana-Champaign
1409 West Green Street
Urbana, Illinois 61801

Seminar on Stochastic Processes, 1982
Birkhäuser, Boston, 1983

REMARKS ON THE CONVEX MINORANT OF BROWNIAN MOTION*

by

J. W. PITMAN

1. Introduction

Recently Groeneboom [1] studied the concave majorant process of a Brownian motion $(B_t, t \geq 0)$. The purpose of this note is to take a fresh look at some of Groeneboom's results in the context of path decompositions of Williams [7], and to give a simple new description of this concave majorant process.

It turns out to be more convenient to work with the convex minorant process (C_t) of (B_t). Of course, $C_t = -\hat{C}_t$ where $\hat{C}_t$ is the concave majorant of the Brownian motion $(-B_t)$. Let $V \subset (0,\infty)$ be the random set of *vertex times* for (C_t): formally, V is the set of points of increase of the right derivative of (C_t). The simplicity of Figure 1 is justified by Groeneboom's observation that, with probability one, for every $0 < s < t < \infty$, V has a finite number of points in the interval (s,t), and a countably infinite number of points in each of the intervals $(0,s)$ and (t,∞).

To find a point in V, fix $b \in (-\infty,0)$, and consider, as in Figure 1, the unique line of slope b that is tangent to (C_t). Let

---

*Research supported by NSF Grant No. MCS 82-02552.

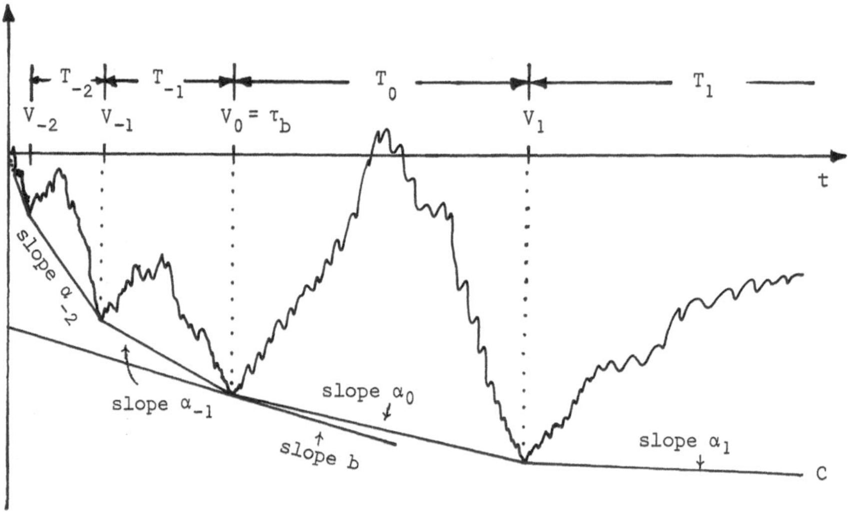

Figure 1. Convex minorant of B is C. The vertex set consists of the points $\ldots, V_{-2}, V_{-1}, V_0, V_1, V_2, \ldots$

τ_b be the last time this line touches (C_t). With probability one we can now define random times V_i indexed by i in the set $\mathbf{Z}$ of integers: $V_0 = \tau_b$; for $n > 0$, V_n is the time of the nth vertex after V_0, and V_{-n} is the time of the nth vertex before V_0. So $V = \{V_i, i \in \mathbf{Z}\}$ a.s.

For $i \in \mathbf{Z}$ define

$$T_i = V_{i+1} - V_i$$

$$\alpha_i = (C_{V_{i+1}} - C_{V_i})/T_i.$$

So α_i is the slope of the ith linear segment of the convex minorant, whose length is T_i. Note that α_i increases from $-\infty$ to 0 as i runs through $\mathbb{Z}$. Since

$$V_i = \sum_{j<i} T_j, \qquad C(V_i) = \sum_{j<i} \alpha_j T_j,$$

and the trajectory of the convex minorant is a succession of line segments between the points (V_i, C_{V_i}), $i \in \mathbb{Z}$, the whole process (C_t) is determined by the random variables (T_i, α_i), $i \in \mathbb{Z}$. The joint distribution of these random variables is given by the following theorem, whose proof will be sketched in Section 2.

(1.1) THEOREM

(i) α_0 *is uniformly distributed on* $(b,0)$*, and conditional on* $\{\alpha_0,\ldots,\alpha_n\}$*,* α_{n+1} *is uniformly distributed on* $(\alpha_n, 0)$;

(ii) α_{-1} *has density* bx^{-2} *on the interval* $(-\infty,b)$*, and conditional on* $\{\alpha_{-1},\ldots,\alpha_{-n}\}$*,* α_{-n-1} *has density* $\alpha_{-n}x^{-2}$ *on the interval* $(-\infty,\alpha_{-n})$;

(iii) *the sequences* $(\alpha_n, n \geq 0)$ *and* $(\alpha_{-n}, n > 0)$ *are independent.*

(iv) *Conditional on all the slopes* (α_i)*, the lengths* T_i *of the segments with these slopes are independent, and* T_i *has a Gamma* $(\frac{1}{2}, \frac{1}{2}\alpha_i^2)$ *distribution* :

$$P(T_i \in dt \mid \alpha_i = a, \alpha_j \text{ for } j \neq i) = a(2\pi t)^{-\frac{1}{2}} e^{-\frac{1}{2}a^2 t}\, dt.$$

This theorem is closely related to the result of Groeneboom [1] that the process

$$(\tau_a, -\infty < a < 0) \tag{1.2}$$

has independent increments. It was observed by R. Blumenthal, and brought to my attention by R. Getoor, that this result of Groeneboom's is an immediate consequence of the path decomposition of Williams [7], Theorem 2.1. Indeed, for $b < 0$, τ_b is the last time that the positively drifting Brownian motion $(B_t - bt)$ touches its overall minimum. According to Williams' path decomposition, the processes

$$(1.3) \qquad (B_t, 0 \le t \le \tau_b) \text{ and } (B_{\tau_b + s} - B_{\tau_b} - bs, 0 \le s < \infty)$$

are independent. For $-\infty < a \le b$, τ_a is a function of the first process in (1.3), while for $b \le c < 0$, $\tau_c - \tau_b$ is a function of the second process in (1.3), whence the independent increments property of the process $(\tau_a, -\infty < a < 0)$. The key to Theorem (1.1) is Williams' description of the second process in (1.3) as a diffusion process which can be identified as a three-dimensional Bessel process with drift $|b|$. See Section 2 for details.

The variables (α_i, T_i), $i \in \mathbf{Z}$, of Theorem (1.1) are the times and magnitudes of the jumps of Groeneboom's independent increments process $(\tau_a, a < 0)$. Thus, by the standard theory of increasing processes with independent increments, the points (α_i, T_i), $i \in \mathbf{Z}$, form a Poisson point process. A simple formula for the expectation measure μ of this Poisson point process on $(-\infty,0) \times (0,\infty)$ can be deduced at once from Theorem (1.1):

$$(1.4) \qquad \mu(da \times dt) = (2\pi t)^{-\frac{1}{2}} e^{-\frac{1}{2}a^2 t}\, da\, dt, \quad a < 0 < t.$$

In particular, parts (i), (ii), and (iii) of Theorem (1.1) amount to the formula

$$(1.5) \qquad \mu(da \times (0,\infty)) = |a|^{-1}\, da, \quad a < 0$$

which is just Groeneboom's result that the number of linear segments in the convex minorant with slopes in the interval (b,c) is a Poisson random variable with parameter

$$\int_b^c |a|^{-1}\, da = \log(b/c).$$

Groeneboom also gives formulae for the densities of increments of the process (τ_a), as well as for the density of C_t, in terms of certain Gaussian integrals. These could be derived from the formula (1.4) by inverting the Laplace transform in the usual Lévy-Khintchine representation.

The argument for Theorem (1.1) also yields Groeneboom's striking description of the Brownian motion conditional on the convex minorant process. Here is an informal statement of his result:

(1.6) THEOREM. (Groeneboom). *Conditional on the convex minorant process* $(C_t, t \geq 0)$, *the process* $(B_t - C_t, t > 0)$ *is a succession of independent Brownian excursions between the vertices of* (C_t).

It would be most interesting to find a way of casting this result into the general Markovian excursion theory of Maisonneuve [2]. The argument presented here makes essential use of the special invariance of Brownian motion under time inversion, and consequent properties of Bessel processes and their bridges.

2. Calculations with Bessel processes

A three-dimensional Bessel process starting at 0 with drift parameter $d \geq 0$, or $BES^0(3,d)$, is a Markov process with continuous paths identical in law to the radial part of a three-dimensional Brownian motion started at the origin with constant drift d in some fixed direction. See Rogers and Pitman [4] for a detailed account of this process.

Let X be the second process in the decomposition (1.3), that is,

$$X_s = B(\tau_b + s) - B(\tau_b) - bs,$$

where $b < 0$ is fixed. According to Williams [7], X is a $BES^0(3,d)$ for $d = |b|$. Now put

$$T = T_0 = \inf\{s > 0: B(\tau_b + s) = C(\tau_b + s)\},$$

the interval between time τ_b and the next vertex of the convex minorant (see Figure 1). In terms of X (see Figure 2),

$$T = \inf\{s > 0: X_s/s = \inf_{0 < v < \infty} X_v/v\} \tag{2.1}$$

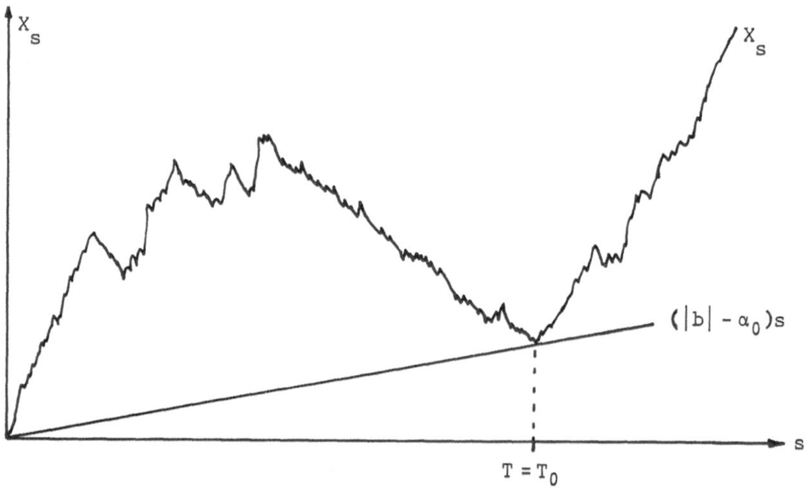

Figure 2. T is the first time (X_s/s) touches its minimum.

Each of Theorems (1.1) and (1.6) is, in view of these remarks, a consequence of repeated applications of the following decomposition of a $BES^0(3,d)$, which is simply the transformation by time inversion of Williams' decomposition at the minimum of a $BES^d(3,0)$, that is, an ordinary three-dimensional Bessel process started at d.

(2.2) THEOREM. *Let* X *be a* $BES^0(3,d)$, *with* $d > 0$, *let* T *be defined by* (2.1), *and put* $\gamma = X_T/T$. *Then,*

(i) $0 < T < \infty$ a.s.;

(ii) γ *is uniformly distributed on* $[0,d]$;

(iii) *conditional on* $\gamma = g$, T *has a Gamma* $(\frac{1}{2},\frac{1}{2}(d-g)^2)$ *distribution:*

$$P(T \in dt \mid \gamma = g) = dt(d-g)(2\pi t)^{-\frac{1}{2}} e^{-\frac{1}{2}(d-g)^2 t}.$$

(iv) *Conditional on* $\gamma = g$ *and* $T = t$, *the processes*

$$(X_s - \gamma s,\ 0 \le s \le T) \quad and \quad (X_{T+v} - X_T - \gamma v,\ v \ge 0)$$

are independent; the first of these processes is a Brownian excursion of length t, *and the second is a* $BES^0(3,d-g)$.

PROOF. Let $Y_u = u\,X(1/u)$, $u > 0$, and put $Y_0 = d$. The familiar time inversion property of Brownian motion implies that Y is a $BES^d(3,0)$, that is, a continuous Markov process identical in law to the radial part of a three-dimensional Brownian motion with no drift, started at distance d from the origin. (See also Watanabe [5], Pitman and Yor [3].) But in terms of the $BES^d(3,0)$ process Y, $1/T = S$ where

$$S = \sup\{u > 0 : Y_u = \inf_{0<t<\infty} Y_t\}$$

is the last time Y attains its overall minimum, and γ is the value of this minimum. According to Williams [7], Theorem (3.1),

(2.3) (a) $0 < S < \infty$ a.s.

(b) γ is uniformly distributed on $[0,d]$.

(c) Conditional on $\gamma = g$, the processes

$$(Y_{S+u} - \gamma,\ u \geq 0) \quad \text{and} \quad (Y_t,\ 0 \leq t \leq S)$$

are independent; the first of these processes is a $BES^0(3,0)$, and the second is a Brownian motion started at d and run until it first hits g.

Now part (i) of the Theorem results from (a), while (ii) is just (b). To obtain (iii), observe from (c) that conditional on $\gamma = g$ the variable $S = 1/T$ has the same distribution as the time it takes Brownian motion to hit a point at distance $(d-g)$ from its origin, and use the well known formula for this hitting time density. Finally, (iv) too can be disentangled from (c). Here, for example, is how the Brownian excursion appears. Let $Z_0 = X_s - \gamma s$, $0 \leq s \leq T$, and let

$$R_u = Y_{S+u} - \gamma, \quad u \geq 0,$$

so (R_u) is the $BES^0(3,0)$ in (c). A little algebra reveals that

$$Z_{T-s} = \frac{T-s}{T} R^*\left(\frac{Ts}{T-s}\right), \quad 0 < s < T, \tag{1.4}$$

where $R^*(v) = T\,R(v/T^2)$ is a $BES^0(3,0)$ by Brownian scaling. But given $T = t$ the process in (1.4) is, as noted by Williams [6], a Brownian excursion of length t, hence so is Z by time reversal. (See also Pitman and Yor [3], Theorem 5.8, which identifies Z as a 3-dimensional Bessel bridge from 0 to 0 over time t, and explains the connection between time inversion and the transformation involved in (1.4).) The remaining assertions of (iv) can be derived similarly.

References

1. P. GROENEBOOM. The concave majorant of Brownian motion. Technical Report No. 6, Dept. Statistics, Univ. of Washington, Seattle. To appear in *Ann. Probab.*

2. B. MAISONNEUVE. Exit systems. *Ann. Probab.* *3* (1975), 399-411.

3. J.W. PITMAN and M. YOR. Bessel processes and infinitely divisible laws. *Stochastic Integrals*, ed. D. Williams, 285-292. Lecture Notes in Math. *851*. Springer-Verlag, Berlin, 1981.

4. L.C.G. ROGERS and J.W. PITMAN. Markov functions. *Ann. Probab.* *9* (1981), 573-582.

5. S. WATANABE. On time inversion of one-dimensional diffusion processes. *Z. Wahrscheinlichkeitstheorie verw. Gebiete* *31* (1975), 115-124.

6. D. WILLIAMS. Decomposing the Brownian path. *Bull. Amer. Math. Soc.* *76* (1970), 871-873.

7. D. WILLIAMS. Path decomposition and continuity of local time for one-dimensional diffusions, I. *Proc. London Math. Soc., Ser. 3, 28* (1974), 738-768.

J. W. PITMAN
Department of Statistics
Univ. of California at Berkeley
Berkeley, CA 94720

Seminar on Stochastic Processes, 1982
Birkhäuser, Boston, 1983

REMARKS ON ENERGY

by

Z.R. POP-STOJANOVIC and K. MURALI RAO

1. Introduction

This paper contains a few remarks on results given in [2] and [4]. In [4], the authors have obtained some results on energy by using such probabilistic tools as sub-Markov resolvents, Revuz measures [5], and additive functionals. These tools are more general than Dirichlet space techniques and kernel theory customarily used when dealing with the concept of energy.

Throughout this paper, $X = (\Omega, F, F_t, X_t, \theta_t, P^x)$ will denote a transient Hunt process with locally compact state space $(E, \mathcal{E})$ with a countable base. Transience here means that, for each x and for each compact set K,

$$P^x[\gamma_K < \infty] = 1,$$

where γ_K is the last exit time from K, that is, $\gamma_K = \sup\{t: X_t \in K\}$. This definition of transience is equivalent to saying that the excessive function $P_K 1$ is purely excessive for each company set K. For all concepts and notations used in the sequel we refer to Blumenthal and Getoor [1].

We assume that there is an excessive reference measure, which is a Radon measure, and which will be denoted by dx. Further, we assume that points are polar sets.

Under these assumptions, it is always possible to choose the densities $u^\alpha(x,y)$ of $U^\alpha(x,\cdot)$, $\alpha \geq 0$, such that

$$u(x,y) = u^\alpha(x,y) + \alpha U^\alpha u(x,y)$$

for all x, y. In particular, the "duals"

$$\hat{U}^\alpha f(y) = \int u^\alpha(x,y) f(x) dx, \quad \alpha \geq 0, \quad f \geq 0,$$

are well-defined. Then, $\hat{U}^\alpha$ is a sub-Markov resolvent.

2. Energy

Our first remark is the following theorem, which is part of a starting hypothesis in [2]. This theorem is useful in justifying the use of Fubini's theorem.

(1) THEOREM. *There is a strictly positive function* a *such that, for almost every* x,

(2) $$0 < Ua(x) \leq 1 \quad \textit{and} \quad 0 < \hat{U}a(x) \leq 1.$$

PROOF. Let (D_n) be a base of relatively compact open sets. For each n, put

$$p_n = P_{D_n} 1, \quad p = \sum_{n=1}^{\infty} 2^{-n} p_n.$$

Then, $p_n = 1$ on D_n. We claim that, for each $t > 0$,

(3) $$P_t p(x) < p(x).$$

Indeed, $P_t p(x) = p(x)$ implies that

(4) $$P_t p_n(x) = p_n(x)$$

for all n. If D_k's are elements of the base of open sets decreasing to the point x, then $p_k(y)$ decreases to zero for each y, because the point set $\{x\}$ is polar and the process is transient and quasi-left continuous. But, this contradicts (4), since $1 = p_k(x) = E^x[p_k(X_t)]$ for each k with $x \in D_k$.

Now the assumption of transience implies that, for each n and x, $\lim_{t\to\infty} P_t p(x) = 0$. The same holds true for p:

(5) $$\lim_{t\to\infty} P_t p(x) = 0.$$

For any $t > 0$, the function $a = (p - P_t p)/t$ is strictly positive and satisfies the relation

$$Ua = \frac{1}{t}\int_0^t P_s p \, ds.$$

Hence, $0 < Ua \leq 1$. By making a smaller if necessary, we may assume that a is integrable. Clearly, then, $\hat{U}a$ is strictly positive almost everywhere. By making a still smaller but strictly positive and by using the maximum principle, it is easy to show that there is $b > 0$ such that $\hat{U}b \leq 1$. Then, the minimum of a and b is the function which meets the requirements of the theorem. □

An immediate consequence of this theorem is the following:

(6) COROLLARY. *For each function* $f \geq 0$ *with* $\int f < \infty$, Uf *is a potential.*

PROOF. Let a be the function introduced in Theorem (1). Then, one has

$$(Uf,a) = (f,\hat{U}a) \leq \int f < \infty. \qquad \square$$

Now we can introduce the following.

(7) DEFINITION. A measurable function f satisfying $(U^\alpha|f|,|f|) < \infty$ is said to be of finite α-energy. Its α-energy is the quantity $e^\alpha(f)$ defined by

$$[e^\alpha(f)]^2 = (U^\alpha f,f), \quad \alpha \geq 0. \qquad \square$$

For $\alpha > 0$, it was shown in [4] that $e^\alpha(f) \geq 0$ and that $e^\alpha(f) = 0$ only if $f = 0$. For $\alpha = 0$, the question was left unanswered in [4]. The following example shows that, for $\alpha = 0$, this statement is not true in general.

(8) EXAMPLE. Let X be the uniform translation to the right; then

$$Uf(s) = \int_0^\infty f(s+t)dt, \quad s \geq 0.$$

So, when f has finite energy, that is, if

$$\int_0^\infty \int_0^\infty |f(s)\ f(t+s)|dt\ ds < \infty,$$

one has

$$(Uf,f) = \int_0^\infty \int_0^\infty f(s)\ f(t+s)dt\ ds = \int_0^\infty f(s)ds \int_s^\infty f(t)dt = \tfrac{1}{2}(\int_0^\infty f(s)ds)^2.$$

It follows that $\int f = 0$ if and only if $(Uf,f) = 0$. □

This example motivates the following.

(9) PROPOSITION. *Let* f *be a signed function such that*

(10) $$(U|f|,|f|) < \infty.$$

Then, $(Uf,f) \geq 0$ *and vanishes only if* $\int_0^\infty f(X_t)dt$ *is* F_0*-measurable.*

PROOF. By using (10) and the function a from Theorem (1), one concludes that

(11) $$U(|f|\cdot U|f|) < \infty$$

almost everywhere. A simple computation gives

$$2U(|f|\cdot U|f|)(x) = E^x[(\int_0^\infty |f(X_t)|dt)^2].$$

Therefore, for each point x at which (11) holds,

(12) $$s(X_t) + \int_0^t f(X_u)du$$

is a martingale relative to (F_t, P^x), where

$$s(x) = E^x[A_\infty], \quad A_\infty = \int_0^\infty f(X_t)dt.$$

Define $t(x) = E^x[(A_\infty - s(X_0))^2]$ where (11) holds and put $t(x) = \infty$ elsewhere. Using (12), a simple computation shows that $t(x)$ is super-averaging. Denote by the same symbol its excessive regularization. Then, on the set (11) holds, we have $2U(fUf) = E^x[A_\infty^2] \geq t(x)$ for almost every x. Let ν be the Revuz [5] measure of $t(x)$. Then, the last inequality implies that

$$\nu(1) \leq 2(f,Uf).$$

So, if $(f,Uf) = 0$, then $\nu(1) = 0$, which in turn implies that $t = 0$, that is, $A_\infty = s(X_0)$ as desired. □

A few final remarks are in order here. In the classical case, the set of measures of finite energy is complete in the energy norm. J. Glover [3] shows that the same conclusion is valid under duality and Hypothesis (H) of Hunt. We can re-formulate the obvious result to say that the set of regular potentials is complete in the energy norm. This formulation leads to generalizations which will be the subject of a later publication.

References

1. R.M. BLUMENTHAL and R.K. GETOOR. *Markov Processes and Potential Theory*. Academic Press, New York, 1968.

2. R.K. GETOOR and M.J. SHARPE. Some remarks on energy and duality. To appear.

3. J. GLOVER. Energy and the maximum principle for nonsymmetric Hunt processes. *Theory of Probability and its Applications, 26* (1981), 757-768.

4. Z.R. POP-STOJANOVIC and K.M. RAO. Some results on energy. *Seminar on Stochastic Processes, 1981,* 135-150. Birkhäuser, Boston, 1981.

5. D. REVUZ. Mesures associées à fonctionnelles additive de Markov. *Trans. Amer. Math. Soc. 148* (1970), 501-531.

Z.R. POP-STOJANOVIC
Department of Mathematics
University of Florida
Gainesville, Florida 32611

K. MURALI RAO
Department of Mathematics
University of Florida
Gainesville, Florida 32611

Seminar on Stochastic Processes, 1982
Birkhäuser, Boston, 1983

STOCHASTIC INTEGRATION WITH RESPECT TO LOCAL TIME

by

JOHN B. WALSH

§1. Introduction

Let $\{B_t: t \geq 0\}$ be a standard Brownian motion with $B_0 = 0$, defined on a probability space (Ω, F, P). Define the occupation time below x by

$$H_t^x = \int_0^t I_{\{B_s \leq x\}} ds$$

and let L_t^x be the local time at x:

$$L_t^x = \frac{d}{dx} H_t^x .$$

Let $\rho(t,x)$ be the inverse of H:

$$\rho(t,x) = \inf\{s: H_s^x > t\}$$

and define the time-changed process Y_t^x by

$$Y_t^x = B_{\rho(t,x)}.$$

We define the *excursion fields* $\{E_x, x \in \mathbb{R}\}$ by

$$E_x^0 = \sigma\{Y_t^x, t \geq 0\},$$

and let E_x be E_x^0 augmented by all null-sets of F.

The excursion fields were introduced in [3]. It was shown there that they formed a right-continuous, increasing filtration, which is in a sense richer than the usual filtration $F_t = \sigma\{B_s: s \leq t\}$, but that the two fields shared many properties. It was conjectured that all martingales relative to the excursion fields were continuous (in [3], there was a possibility of a single jump at the process minimum, but in the case we consider here, there is no minimum).

It was further conjectured that there might be an integral representation theorem behind this, and that any square-integrable martingale $\{M_x, E_x, x \in \mathbb{R}\}$ should have a representation as a stochastic integral with respect to local times. David Williams [4] has established the first conjecture by finding a set of continuous martingales which is dense [4]. One of the aims of this note is to establish the second, by showing that there exists a constant $M_{-\infty}$ and process $\{\phi(t,x): t \geq 0, x \in \mathbb{R}\}$ such that

$$M_x = M_{-\infty} + \iint_{\mathbb{R}_+ \times (-\infty,x]} \phi(t,x)dL_t^x. \tag{1.1}$$

Local time L_t^x, being a two-parameter process, can be used to define a stochastic integral over the plane as above. It can also be used to define a Stieltjes integral over t for fixed x, and to define a stochastic integral over x for fixed t. This gives three possible types of integrals with respect to local time. There are evidently many relations among these integrals. We shall look at a few of these in the latter part of the paper, but the first thing we must do is to define the stochastic area integral in (1.1).

A few words of explanation are in order here. We always assume $B_0 = 0$, and much of our analysis is carried out by looking at the excursions from a point x. If $x < 0$, the initial excursion $\{B_t, 0 \le t \le T_x\}$ is above x, and hence is not $\mathcal{E}_x$-measurable, while if $x > 0$, it is $\mathcal{E}_x$-measurable. This initial excursion is rather special, which is why we shall treat the cases $x < 0$ and $x \ge 0$ separately so often in what follows.

§2. Stochastic Integrals

We will define an area integral with respect to the local time L_t^x. This is somewhat similar to the integral of weakly-adapted integrands of [1], but the hypotheses used there are not satisfied here, so we must retrace the construction.

Regard L_t^x as a two-parameter process. It is jointly continuous in t and x, and it generates an additive set function L on the rectangles of the right half-plane by

$$L((s,t] \times (x,y]) = L_t^y - L_t^x - L_s^y + L_s^x. \tag{2.1}$$

We are going to switch the roles of space and time, treating x as if it were the new time variable, and t as if it were a space variable. We will use L to define integrals over the right half-plane, but we must be careful. We won't use (2.1) on all rectangles, but only on a certain class of special rectangles. We recall the following definition from [3, p. 168].

DEFINITION. A random variable $\tau \ge 0$ is $\mathcal{E}_x$-*identifiable* if $\{\tau = 0\}$, L_τ^x, and $\tau - \sup\{s < \tau: B_s = x\}$ are all $\mathcal{E}_x$-measurable.

Intuitively, an $\mathcal{E}_x$-identifiable random variable is an instant of time which can be identified by looking only at the excursions below x. One example would be the first hitting time T_y of any point $y \le x$. This intuition is more obvious in the following characterization than in the definition.

PROPOSITION 2.1. A necessary and sufficient condition that a positive random variable τ be $\mathcal{E}_x$-identifiable is that the following three conditions hold:

(i) $\{\tau = 0\} \in \mathcal{E}_x$;

(ii) H^x_τ is $\mathcal{E}_x$-measurable;

(iii) $B_\tau \le x$ a.s. on $\{\tau > 0\}$.

Before proving this, we first establish a fundamental property of identifiable random variables.

LEMMA 2.2. Let τ be an $\mathcal{E}_x$-identifiable time. Then, on the set $\{\tau > T_{(-\infty,x]}\}$, if t is such that $L^x_t = L^x_\tau$, it follows that $B_t \le x$. In particular, τ does not coincide with the initial or final point of any excursion above x.

PROOF. Let $\gamma = \sup\{s < \tau: B_s = x\}$, and set $h = L^x_\gamma = L^x_\tau$. Then h is $\mathcal{E}_x$-measurable, and is thus independent of the point process $\{\Xi^+_x(t), t \ge 0\}$ of excursions above x which was defined in [3, pp. 162, 166]. Thus $P\{\Xi^+_x(h) = \partial\} = 1$, i.e. h is not an excursion time for Ξ^+_x. In terms of B, this means that γ is not the initial time of some excursion of B above x. But if $\tau > \gamma$, γ is the initial point of some excursion, hence, of an excursion below x. Since τ falls during this excursion, $B_\tau \le x$. If $\tau = \gamma$, $B_\tau = x$. But now, if $L_t^x = L_\tau^x$, then τ must fall during the same excursion, hence $B_t \le x$ as well.

PROOF (of Prop. 2.1). If τ is identifiable, (i) follows from the definition, and (iii) follows from Lemma 2.2. To see (ii), let $\gamma = \sup\{s < \tau: B_s = x\}$, $h = L_\tau^x$, and note that

$$H_\gamma^x = \inf\{u: L_{\rho(u,x)}^x = h\};$$

as h and $L_{\rho(u,x)}^x$ are E_x-measurable, so is H_γ^x. By Lemma 2.2, $B_t \le x$ if $\gamma \le t \le \tau$, and $H_\tau^x = H_\gamma^x + (\tau - \gamma)$, which is E_x-measurable since $\tau - \gamma$ is.

Conversely, suppose (i) - (iii) hold. Then L_τ^x is E_x-measurable, since

$$L_\tau^x = \lim \frac{1}{\varepsilon}(H_\tau^x - H_\tau^{x-\varepsilon}).$$

Finally, $H_\gamma^x = \inf\{u: L_{\rho(u,x)}^x = h\}$ is E_x-measurable; since $B_t < x$ on (γ,τ), $\tau - \gamma = H_\tau^x - H_\gamma^x$, which is E_x-measurable, and so τ is E_x-identifiable.

Here are some easy consequences of Proposition 2.1. We will omit the proofs.

COROLLARY 2.3. (i) If τ is E_x-identifiable, it is also E_y-identifiable for any $y \ge x$.

(ii) If τ_n, $n = 1,2,\ldots$ are E_x-identifiable and converge to a limit τ, then τ is E_x-identifiable.

(iii) If σ and τ are E_x-identifiable, so are $\sigma \wedge \tau$ and $\sigma \vee \tau$.

(iv) If τ is E_x-identifiable, B_τ is E_x-measurable.

(v) If T is E_x-measurable, $\rho(T,x)$ is E_x-identifiable.

Notice that we do not claim that τ itself is E_x-measurable. It usually is not. A constant time, other than zero, is not E_x-identifiable. In fact, it is not hard to show that if $x < 0$ and τ is E_x-identifiable, then $P\{\tau = t\} = 0$ for any $t > 0$.

PROPOSITION 2.4. $\bigcap_x E_x$ is trivial.

PROOF. E_x decreases as x decreases, so $\bigcap_x E_x = \bigcap_{x<0} E_x$. Let $S_x = \inf\{s > 0: H_{T_x+s} > t\}$ and note that for $x < 0$, $\rho(t,x) = T_x + S_x$. Now S_x is measurable with respect to $\sigma\{B_{T_x+u}, u \geq 0\}$, hence so is $B_{\rho(t,x)}$. Thus $E_x \subset \sigma\{B_{T_x+u}, u \geq 0\}$ for all $x < 0$, so $\bigcap_x E_x \subset \bigcap_{x<0} \sigma\{B_{T_x+u}, u \geq 0\}$. Now suppose $A \in \bigcap_x E_x$. Then, for each $x < 0$, the strong Markov property implies that A is independent of F_{T_x} given B_{T_x}. But $B_{T_x} \equiv x$, so that A is independent of F_{T_x}. Thus A is independent of $\bigvee_x F_{T_x} = F$. In particular, A is independent of itself, so $P\{A\} = 0$ or 1.

DEFINITION. A pair $\sigma \leq \tau$ of E_x-identifiable times is *proper* if $\sigma > 0$ a.s. and $E\{(L_\tau^x - L_\sigma^x)^2\} < \infty$.

REMARKS. 1° Let $\tau_n = \rho(n,x)$. Then τ_n is E_x-identifiable and $L_{\tau_n}^x \equiv n$, so if $0 < \sigma \leq \tau_n$ is identifiable, $\{\sigma,\tau_n\}$ is a proper pair. If $\sigma \leq \tau$ is any pair of E_x-identifiable times for which $\sigma > 0$, we may replace them by $\sigma \wedge \tau_n$ and $\tau \wedge \tau_n$ respectively to get a proper pair for which $L_\tau^x - L_\sigma^x$ is even bounded. We will shortly see (Theorem 2.5) that if $\sigma \leq \tau$ are proper E_x-identifiable times, they are also proper E_y-identifiable times for any $y > x$.

2° The reason for insisting that $\sigma > 0$ in the definition is simply to assure that $B_\sigma \leq x$ (Prop. 2.1). For $x > 0$, $B_\sigma < x$ on $\{\sigma = 0\}$ as well as on $\{\sigma > 0\}$, so one could easily dispense with the condition there. This would make no difference to the definitions below, however.

DEFINITION. A *special* (stochastic) *rectangle with base* x_0 is a subset K of $\mathbb{R}_+ \times \mathbb{R} \times \Omega$ of the form

$$K = \{(s,x,\omega): \sigma(\omega) < s \le \tau(\omega),\ x_0 < x \le x_1\}, \tag{2.2}$$

where $x_0 < x_1$ and $0 < \sigma \le \tau$ is a proper pair of $\mathcal{E}_{x_0}$-identifiable times.

DEFINITION. The σ-field $\mathcal{I}$ of subsets of $\mathbb{R}_+ \times \mathbb{R} \times \Omega$ generated by the special rectangles is called the σ-field of *identifiable sets*, or simply, the *identifiable* σ-field.

DEFINITION. A process $\phi = \{\phi(t,x),\ t \ge 0,\ x \in \mathbb{R}\}$ is *identifiable* if it is $\mathcal{I}$-measurable as a function of (t,x,ω).

REMARKS. 1° $\mathcal{I}$ is also generated by all sets of the form $(\sigma,\tau] \times (x_0,x_1]$, where $0 < \sigma \le \tau$ are identifiable, but not proper. This follows since such a set is a union of special rectangles of the form $(\sigma \wedge \tau_n,\ \tau \wedge \tau_n] \times (x_0,x_1]$.

2° $\mathcal{I}$ also contains rectangles of the form $[0,\sigma] \times (x_0,x_1]$, where $\sigma > 0$ is identifiable.

3° If $\sigma > 0$ is identifiable, then $B_\sigma \le x$ (Prop. 2.1) so that if $x < 0$, $\sigma \ge T_x$. It follows that if $A \in \mathcal{I}$ and if for some ω, $(0,x,\omega) \in A$ then the set $[0,T_x] \times \{x\} \times \{\omega\} \subset A$. (This is true for special rectangles and their complements, etc.) Consequently, if ϕ is identifiable and $x < 0$,

$$\phi(0,x) = \phi(T_x,x).$$

We are going to construct stochastic integrals of identifiable processes. This is a standard construction for the most part. When it differs from the familiar constructions of stochastic area integrals, it is usually because of the special properties of identifiable sets and processes. In order to keep from overloading the exposition, we have put a number of results on identifiability in the appendix.

Let us recall two results from [3] which link local time and identifiability. These are the basic properties we need to construct our integrals.

THEOREM 2.5. Let $\sigma \le \tau$ be a proper pair of $\mathcal{E}_{x_0}$-identifiable times, and define $M_x = L^x_\tau - L^x_\sigma$. Then $\{M_x, \mathcal{E}_x, x \ge x_0\}$ is a positive continuous square-integrable martingale with associated increasing process

$$\langle M\rangle_x = 4\int_{x_0}^{x} M_y dy.$$

THEOREM 2.6. Let $\sigma_1 \le \tau_1 \le \sigma_2 \le \tau_2$ be proper pairs of $\mathcal{E}_x$-identifiable times, and put $M^i_x = L^x_{\tau_i} - L^x_{\sigma_i}$. Then the martingales $\{M^1_x, \mathcal{E}_x, x \ge x_0\}$ and $\{M^2_x, \mathcal{E}_x, x \ge x_0\}$ are orthogonal.

DEFINITION. A function $\phi = \phi(t,x,\omega)$ is a *simple function* if it is a finite sum of functions of the form XI_R, where R is a special rectangle and X is a bounded $\mathcal{E}_{x_0}$-measurable random variable, where x_0 is the base of R.

We construct the stochastic integral of a simple function. Let $R = (\sigma,\tau] \times (x,y]$ be a special rectangle, and let X be bounded and $\mathcal{E}_x$-measurable. We define

$$\iint \phi dL = X(L^y_\tau - L^y_\sigma - L^x_\tau + L^x_\sigma). \tag{2.3}$$

We then extend $\iint \phi dL$ to simple functions ϕ by linearity.

It is easily verified that this integral is independent of the particular representation of ϕ as a sum of indicator functions. The main thing to verify is that the intersection of two special rectangles is a special rectangle, which follows from Corollary 2.3 (i) and (iii).

Let ϕ be a simple function and define a process $\{\phi\cdot L(x),\ x \in \mathbb{R}\}$ by

$$\phi\cdot L(x) = \iint \phi(t,y) I_{\{y \le x\}} dL. \tag{2.4}$$

THEOREM 2.7. Let ϕ and ψ be simple functions. Then

(i) $\{\phi\cdot L(x),\ \mathcal{E}_x,\ x \in \mathbb{R}\}$ is a continuous square-integrable martingale;

(ii) $E\{(\phi\cdot L(x))^2\} = 4E\Big\{\int_0^\infty \phi(s,B_s)^2 I_{\{B_s \le x\}} ds\Big\}$;

(iii) $E\{\phi\cdot L(x)\psi\cdot L(x)\} = 4E\Big\{\int_0^\infty \phi(s,B_s)\psi(s,B_s) I_{\{B_s \le x\}} ds\Big\}$.

PROOF. Let $K = (\sigma,\tau] \times (x_0,y_0]$ be a special rectangle and put $M_x = L_\tau^x - L_\sigma^x$, $x \ge x_0$. If X is bounded and $\mathcal{E}_{x_0}$-measurable and $\phi = XI_K$, then

$$\phi\cdot L(x) = \begin{cases} 0 & \text{if } x < x_0 \\ X(M_x - M_{x_0}) & \text{if } x_0 \le x \le y_0. \\ X(M_{y_0} - M_{x_0}) & \text{if } x > y_0 \end{cases}$$

This is a continuous square-integrable martingale by Theorem 2.5 above and, if $x_0 \le x \le y_0$,

$$E\big\{(\phi\cdot L(x))^2 \,|\, \mathcal{E}_{x_0}\big\} = X^2 E\{(M_x - M_{x_0})^2\}$$

$$= 4X^2 E\{\int_{x_0}^{x} M_y dy\}$$

$$= 4X^2 E\{\int_{x_0}^{x} L^y_\tau dy - \int_{x_0}^{x} L^y_\sigma dy\}.$$

Since the local time is the derivative of the occupation time, this is

$$= 4X^2 E\{\int_0^\tau I_{(x_0,x]}(B_s)ds - \int_0^\sigma I_{(x_0,x]}(B_s)ds\}$$

$$= 4E\{\int_0^\infty \phi^2(s,B_s) I_{\{B_s \le x\}} ds\}.$$

Next, let K_1 and K_2 be disjoint special rectangles: $K_i = (\sigma_i,\tau_i] \times (x_i,y_i]$, $i = 1,2$. Suppose first $y_1 \le x_2$. Then $I_{K_1} \cdot L(x)$ is $\mathcal{E}_{x_2}$-measurable for any x while $E\{I_{K_2} \cdot L(x) | \mathcal{E}_{x_2}\} = 0$, hence

$$E\{(I_{K_1} \cdot L(x))(I_{K_2} \cdot L(x))\} = 0. \tag{2.5}$$

Next, suppose $\tau_1 \le \sigma_2$ and put $M^i_x = L^x_{\tau_i} - L^x_{\sigma_i}$, $i = 1,2$. By Theorem 2.6, $\{M^1_x, x \ge x_1 \vee x_2\}$ and $\{M^2_x, x \ge x_1 \vee x_2\}$ are orthogonal martingales, and hence (2.5) holds. The general case follows since if K_1 and K_2 are disjoint special rectangles, they can be broken into sub-rectangles related in the above ways.

This shows that if $\phi = X_1 I_{R_1}$ and $\psi = X_2 I_{R_2}$, then (ii) holds and, if $R_2 \cap R_1$ is empty, (iii) holds. If, now, ϕ and ψ are simple, write $\psi = \sum X_j I_{R_j}$ and $\psi = \sum Y_j I_{Q_j}$. We may assume that each pair R_j, Q_k of rectangles is either disjoint or identical. Otherwise said, we may assume that there are disjoint special rectangles $R_1,\dots,R_n$, and random variables $X_1,\dots,X_n$ and $Y_1,\dots,Y_n$ such that $\phi = \sum X_i I_{R_i}$, $\psi = \sum Y_j I_{R_j}$. Then

$$E\{(\phi \cdot L)(\psi \cdot L)\} = \sum_i \sum_j E\{X_i Y_j (I_{R_i} \cdot L)(I_{R_j} \cdot L)\}$$

$$= \sum_i E\{X_i Y_i (I_{R_i} \cdot L)(I_{R_i} \cdot L)\}$$

$$= E\{\int_0^\infty \phi(s,B_s)\psi(s,B_s)ds\}.$$

We now extend the integral to more general ϕ.

DEFINITION. If ϕ is identifiable, let $\|\phi\|^2 = 4E\{\int_0^\infty \phi(t,B_t)^2 dt\}$. Let H be the set of identifiable processes for which

(i) $\|\phi\|^2 < \infty$;

(ii) $\phi(0,x) = 0$ for all $x < 0$.

NOTE. The reason for (ii) will become clear later. For the moment let us note that each simple function satisfies it, for a special rectangle does not intersect the x-axis.

Thus the simple functions are in H, and they are dense, so that one can define $\iint \phi dL$ for all $\phi \in H$ by the usual approximation arguments: choose simple ϕ_n such that $\|\phi - \phi_n\| \to 0$, and define $\phi \cdot L = \lim \phi_n \cdot L$. It then follows that:

THEOREM 2.8. Let $\psi, \phi \in H$. Then $\{\phi \cdot L(x), E_x, x \in \mathbb{R}\}$ is a continuous square-integrable martingale with increasing process

$$\langle \phi \cdot L \rangle_x = 4\int_0^\infty \phi(t,B_t)^2 \partial_t H_t^x; \tag{2.6}$$

and

$$\langle \phi \cdot L, \psi \cdot L \rangle_x = 4\int_0^\infty \phi(t,B_t)\psi(t,B_t)\partial_t H_t^x. \tag{2.7}$$

REMARK 1° A well-known property of local time is that for a bounded Borel function f,

$$\int_0^t f(B_s)ds = \int_{-\infty}^\infty f(y)L_t^y dy.$$

This can be extended to functions depending on t as well. If we apply it to (2.6) and (2.7) above, we get

$$(2.8)\quad \begin{cases} \int_0^\infty \phi(t,B_t)^2 I_{\{B_t \le x\}} dt = \int_{-\infty}^{x} \int_0^\infty \phi(t,y)^2 \partial_t L_t^y dy \\ \\ \int_0^\infty \phi(t,B_t)\psi(t,B_t) I_{\{B_t \le x\}} dt = \int_{-\infty}^{x} \int_0^\infty \phi(t,y)\psi(t,y) \partial_t L_t^y dy. \end{cases}$$

2° H is a Hilbert space, isometric to a subspace of $L^2(\Omega)$ under the map $\phi \to \phi \cdot L(\infty)$.

Note that we have not defined the integral of all identifiable functions. Even ignoring integrability conditions, we have condition (ii) of the definition of H. Let us look at this a bit more closely.

We define three sets:

$$S^+ = \{(t,x,\omega): t > \rho(0,x,\omega);\ x \in \mathbb{R}\}$$

$$S^- = \{(t,x,\omega): t \le \rho(0,x,\omega);\ x \le 0\}$$

$$S^0 = \{(t,x,\omega): t = \rho(0,x,\omega);\ x \le 0\}.$$

It is not hard to see that S^+ and S^- are identifiable (see the appendix) but that S^0 is not. Indeed, if K is a special rectangle, $K \cap S^0 = \phi$. In fact, the smallest identifiable set containing S^0 is S^-. Nevertheless, we can compute its "outer" L-measure.

Define $\rho_x = \rho(0,x) = \inf\{t: H_t^x > 0\}$. (Note that $\rho_x = T_x$ a.s. for each given x, but that $\rho_x \not\equiv T_x$.)

Fix n, and let $x_j = -j2^{-n}$. Let K_{nj} be the rectangle

$$K_{nj} = (\rho_{x_j}, \rho_{x_{j+1}}] \times (x_{j+1}, x_j].$$

Now K_{n_j} is not special, since ρ_{x_j} is not $\mathcal{E}_{x_{j+1}}$-identifiable, but we can still define $L(K_{n_j})$ by

$$L(K_{n_j}) = L_{\rho_{x_{j+1}}}^{x_j} - L_{\rho_{x_{j+1}}}^{x_{j+1}} - L_{\rho_{x_j}}^{x_j} + L_{\rho_{x_j}}^{x_{j+1}} = L_{\rho_{x_{j+1}}}^{x_j}$$

since L vanishes at three of the four corners of K_{n_j}. Let $a < b$ and define

$$(2.9) \qquad L\{S^0 \cap \mathbb{R}_+ \times (a,b] \times \Omega\} = \lim_{n\to\infty} \sum_{j:a< x_{j+1} \le b} L(R_{n_j}).$$

Note that $L(K_{n_j}) = L(K_{n_j}^-)$, where $K_{n_j}^- = (0,\rho_{x_{j+1}}] \times (x_{j+1}, x_j]$, so that the same expression would give us

$$L\{S^- \cap \mathbb{R}_+ \times (a,b] \times \Omega\}.$$

PROPOSITION 2.9. With probability one, the above limit exists and equals $2(b-a)$ for all $a < b$.

PROOF. By the strong Markov property applied to the stopping times ρ_{x_j}, we see that the $L_{\rho_{x_{j+1}}}^{x_j}$ are independent and have the same distribution as L_ρ^0, where we let $\rho = \rho_{x_1}$. We know from the general theory of local times that this is exponential. To find its parameter, let $\sigma = \inf\{t:|B_t| = 2^{-n}\}$. Then $\sigma \le \rho$ so that

$$E\{L_\rho^0\} = E\{L_\sigma^0\} + E\{L_\rho^0 - L_\sigma^0 | \sigma < \rho\}P\{\sigma < \rho\}$$
$$= E\{L_\sigma^0\} + \tfrac{1}{2}E^0\{L_\rho^0\},$$

since $P\{\sigma < \rho\} = \frac{1}{2}$ by symmetry and, given $\sigma < \rho$, L_t^0 is constant between σ and the next time it reaches 0, so that $L_\rho^0 - L_\sigma^0$ has the same distribution as L^0_ρ. Now use the fact that $|B_t| - L^0_t$ is a martingale to see that

$$E\{L^0_\rho\} = 2E\{L^0_\sigma\} = 2E\{|B_\sigma|\} = 2^{-n+1}.$$

Now the expectation of the sum in (2.9) is $2(b-a) + O(2^{-n})$ and its variance is $O(2^{-n})$ so that it converges a.s. to its expectation, $2(b-a)$. This holds simultaneously for all rational $a < b \leq 0$ and hence by Dini's theorem (the limit is monotone) for all $a < b \leq 0$, uniformly on compact sets. *Q.E.D.*

This allows us to extend the definition of the integral to identifiable functions which don't satisfy (ii) of the definition of H. Since for any identifiable ϕ, $t \to \phi(t,x)$ is constant on $[0, \rho_x]$, the above suggests that $\int \phi I_{S^-}\, dL = 2\int \phi(0,x)dx$.

DEFINITION. Let $\hat{H}$ be the set of identifiable ϕ for which

(i) $\phi I_{S^+} \in H$;

(ii) $x \to \phi(0,x)$ is integrable on $(-\infty,0]$.

For $\phi \in \hat{H}$, we define

$$(2.10) \qquad \phi \cdot L(x) = (\phi I_{S^+}) \cdot L(x) + 2\int_{-\infty}^{x} \phi(0,y)dy.$$

REMARK. If $\phi \in H$, $\phi \cdot L$ is a martingale. If $\phi \in \hat{H}$, $\phi \cdot L$ is a martingale if $\phi(0,y) = 0$ for a.e. y.

We will extend the integral further by a simple localization. Let ϕ be identifiable and let

$$[\phi]_x = \int_0^\infty \phi(t,B_t)^2 \partial_t H^x_t.$$

Then $[\phi]_x$ is E_x-adapted (example 2° of the appendix) and, if $[\phi]_x < \infty$

all x, $x \to [\phi]_x$ is continuous by the dominated convergence theorem. Suppose

$$(2.11) \qquad [\phi]_x < \infty \quad \text{and} \quad \int_{-\infty}^{0} |\phi(0,x)|dx < \infty \quad \text{a.s.}$$

Let $Z_n = \inf\{x: [\phi]_x \geq n\}$. Then Z_n is a stopping time relative to (E_x), so $\{I_{\{x \leq Z_n\}}, x \geq 0\}$ is E_x-adapted and left-continuous, hence predictable (as a one-parameter process relative to E_x). Thus $\phi_n(t,x) \overset{\text{def}}{=} \phi(t,x)I_{\{x \leq Z_n\}}$ is identifiable (Appendix, Example 3°). Moreover, $[\phi]_{Z_n} = n$ on $\{Z_n < \infty\}$, so $||\phi_n||^2 \leq n$. Thus $\phi_n \in \hat{H}$, so we can define $\phi_n \cdot L$. One can then verify that for $p \geq n$, $\phi_n \cdot L(x) = \phi_p \cdot L(x \wedge Z_n)$. This leads us to define

$$(2.12) \qquad \phi \cdot L(x) = \lim \phi_n \cdot L(x).$$

DEFINITION. Let $\hat{H}_{\ell oc}$ be the set of identifiable processes satisfying (2.11), and let $H_{\ell oc}$ be the set of processes in $\hat{H}_{\ell oc}$ which vanish on S^-.

DEFINITION. If $\phi \in \hat{H}_{\ell oc}$, $\phi \cdot L$ is defined by (2.12).

Then we have the following extension of Theorem 2.8 which, taking (2.10) into account, follows by the usual localization argument using the Z_n defined above.

THEOREM 2.10. If $\phi \in \hat{H}_{\ell oc}$, $\{\phi \cdot L(x) - 2\int_{-\infty}^{x} \phi(0,y)dy, E_x, x > -\infty\}$ is a local martingale. Its quadratic variation is given by

$$(2.13) \qquad \langle \phi \cdot L \rangle_x = 4\int_0^\infty \phi^2(t,B_t)\partial_t H_t^x .$$

If $\psi \in \hat{H}_{\ell oc}$, the quadratic covariation of $\phi \cdot L$ and $\psi \cdot L$ is

$$(2.14) \qquad \langle \phi \cdot L, \psi \cdot L \rangle = 4\int_0^\infty \phi(t,B_t)\psi(t,B_t)\partial_t H_t^x .$$

COROLLARY 2.11. Let ϕ be identifiable, and $a < b$. Then

(i) $E\{\int |\phi(t,a)|\partial_t L_t^b\} \leq 2(a^- - b^-)E\{|\phi(0,a)|\} + E\{\int_0^\infty |\phi(t,a)|\partial_t L_t^a\}.$

(ii) If the right-hand side of (i) is finite,

$$E\{\int_0^\infty \phi(t,a)\partial_t L_t^b | E_a\} = 2(a^- - b^-)\phi(0,a) + \int_0^\infty \phi(t,a)\partial_t L_t^a .$$

PROOF. Define $\hat{\phi}$ by

$$\hat{\phi}(t,y) = \begin{cases} \phi(t,a) & \text{if } y \geq a \\ 0 & \text{if } y < a . \end{cases}$$

Note that $\hat{\phi}$ is also identifiable. If ϕ is the indicator function of a special rectangle K, then clearly

$$(2.15) \qquad \int_0^\infty \phi(t,a)\partial_t L_t^b - \int_0^\infty \phi(t,a)\partial_t L_t^a = \iint_{R_b} \hat{\phi}\, dL.$$

Fix $N > 0$ for the moment and let $A_N = [0,T_{-N}] \times [-N,\infty]$. Let G_N be the class of identifiable set A for which (2.15) holds for the function $\phi = I_{A \cap A_N}$. Then, if $x > -N$, $\int \phi(t,a)\partial_t L_t^x \leq L_{T_{-N}}^x$, which is square-integrable. Now G_N contains special rectangles. It is closed under complements, since

$$\int I_{A_N^c \cap A_N}\, \partial_t L_t^x = L_{T_{-N}}^x - \int I_{A \cap A_N}\, \partial_t L_t^x .$$

It is closed under countable unions, which follows by using monotone convergence on the left-hand side of (2.15) and using (2.12) on the right.

Thus G_N contains $\mathcal{I}$, so if $A \in \mathcal{I}$ and if $\phi = I_{A \cap A_N}$, then $\hat{\phi} \in \hat{H}$ and (2.15) holds. Take the expectation of both sides of (2.15). The right-hand side gives $2(a^- - b^-)E\{\phi(0,a)\}$, which leads to (i). Since ϕ vanishes if $t > T_{-N}$, the two integrals on the left-hand side of (2.15) are bounded by $L^b_{T_{-N}}$ and $L^a_{T_{-N}}$ respectively, both of which are exponential r.v, and thus are integrable. This implies that (ii) holds for this ϕ.

But now if $\phi \geq 0$ is identifiable, it can be written as the increasing limit of functions of the form $\phi_N = \sum_j \alpha_{N_j} I_{A_{N_j}}$ where the α_{N_j} are constants and the A_{N_j} are identifiable subsets of A_N. The corollary holds for each ϕ_N, hence for ϕ by monotone convergence applied to all terms of (i) and (ii). This takes care of positive ϕ, and the general case follows by linearity. *Q.E.D.*

COROLLARY 2.12. Let ϕ be identifiable, let $a < b$, and let $\hat{\phi}(t,x) = \phi(t,a)$ if $a \leq x \leq b$ and $\hat{\phi}(t,x) = 0$ if $x < a$ or $x > b$. If $\hat{\phi} \in \hat{H}_{\ell oc}$ and if $\int |\phi(t,a)| \partial_t L^a_t < \infty$, then

$$\int \phi(t,a)\partial_t L^b_t - \int \phi(t,a)\partial_t L^a_t = \iint_{R_b - R_a} \hat{\phi}(t,y)dL^y_t \; . \tag{2.16}$$

PROOF. By (2.15) and the ensuing argument, (2.16) holds if ϕ is the indicator function of an identifiable subset of A_N for some N. It follows easily that it holds for $\hat{\phi} \in \hat{H}$. If $\hat{\phi} \in \hat{H}_{\ell oc}$ and $\hat{\phi} \geq 0$, let $Z_n = \inf\{x : [\hat{\phi}]_x \geq n\}$ and let $\hat{\phi}_n(t,x) = \hat{\phi}(t,x)$ if $x \leq Z_n$, and set it equal to zero if $x > Z_n$. Then (2.16) holds for $\hat{\phi}_n$, and the right-hand side is just $\hat{\phi}_n \cdot L(b)$. As $n \to \infty$, each of the terms on the left-hand side converges to its limit, since $\hat{\phi}_n(t,b) = \hat{\phi}(t,b)$ on the set $\{Z_n > b\}$, which increases to the whole space. By hypothesis, $\int |\phi(t,a)| \partial_t L^a_t$ is finite hence, a fortiori, so is $\int |\phi(t,a)| \partial_t L^b_t$.

§3. Elementary Integral Formulas

There are several ways to integrate with respect to local time: L_t^x is an increasing function of t for fixed x, so we can define $\int f(t)dL_t^x$ as a Stieltjes integral; it is a semimartingale in x for fixed t [2] so that $\int f(x)dL_t^x$ can be defined as a stochastic integral; and we have just finished defining $\iint f(t,x)dL_t^x$ as a stochastic integral over the plane. There are evidently numerous relations between these integrals. We will discuss several of these in the next sections. In order to distinguish the one-parameter integrals from the double integrals, we will use a round delta for the former, and we will use subscripts to indicate the variables of integration wherever necessary to avoid ambiguity: e.g. $\int f(t)\partial_t L_t^x$ represents the integral over t for fixed x.

We begin with some elementary formulas. Let $K = (S,T] \times (a,b]$ be a special rectangle and let $x \le a$.

Let $J_t = L_t^b - L_t^a - L_S^b + L_S^a$ for $t \ge S$, and let $N_y = L_T^y - L_S^y - L_T^a + L_S^a$ for $y \ge a$. We denote by R_x the rectangle $(0,\infty) \times (-\infty,x]$.

PROPOSITION 3.1.

$$(3.1) \qquad \int_S^T L_t^x \, \partial_t(L_t^b - L_t^a) = \iint_K L_t^x \, dL_t^y ;$$

$$(3.2) \qquad \int_S^T J_t \partial_t L_t^x = \iint_K (L_T^x - L_t^x) dL_t^y ;$$

$$(3.3) \qquad \int_S^T J_t \partial_t J_t = \iint_K N_y dL_t^y + 2 \int_a^b (L_T^y - L_S^y) dy .$$

PROOF. Let $\tau(t) = \inf\{s: L_s^x > t\}$ be the inverse of the local time at x. Let $0 = t_1 < t_2 < \cdots$ where $\lim t_j = \infty$, and let K_j be the special rectangle $K_j = K \cap (\tau(t_j), \tau(t_{j+1})] \times (a,b]$. For each ω, there will be only finitely many j for which K_j is non-empty so that

$$\phi(t,y) \overset{\text{def}}{=} \sum_j t_j I_{K_j}(t,y) \quad\text{and}\quad \phi'(t,y) \overset{\text{def}}{=} \sum_j t_{j+1} I_{K_j}(t,y)$$

are a.s. finite sums. If $\tau(t_j) \le t \le \tau(t_{j+1})$, then $t_j \le L_t^x \le t_{j+1}$, so ϕ and ϕ' converge uniformly to $L_t^x I_K(t,y)$ as the mesh of the partition goes to zero. In particular, the limit is identifiable. Thus

$$\iint_K \phi' dL = \sum_j t_j L(K_j) \le \int_S^T \phi(t,b)\partial_t(L_t^b - L_t^a) \le \sum_j t_{j+1} L(K_j) = \iint_K \phi dL.$$

But both stochastic integrals converge to the right-hand side of (3.1) as the mesh of the partition goes to zero.

Turning to (3.2), write the left-hand side as a limit of Riemann sums:

$$\begin{aligned}\sum_{j=M}^{N} J_{\tau(t_j)}(t_{j+1} - t_j) &= \sum_{j=M}^{N} \sum_{i<j} (t_{j+1} - t_j)L(K_i)\\ &= \sum_{i=M}^{N-1} (t_N - t_{i+1})L(K_i)\end{aligned}$$

where M and N are the minimum and maximum values of j for which K_j is non-empty. We have used the fact that $J_{\tau(t_j)} = \sum_{i=M}^{j-1} L(R_i)$. Now $t_N - t_{i+1} = L^x_{\tau(t_N)} - L^x_{\tau(t_{i+1})}$ so the above expression tends to the right-hand side of (3.2).

The third formula involves Ito's lemma. The left-hand side is just a Stieltjes integral, and equals $\frac{1}{2}J_T^2$. The right-hand side is a stochastic integral with respect to the martingale $\{N_y, y \ge a\}$ (cf. Thm. 2.2). The associated increasing process is

$$\langle N\rangle_z = 4\int_a^z (L_T^y - L_S^y)dy, \quad a \le z \le b.$$

By Ito's Lemma

$$\tfrac{1}{2}N_b^2 = \int_a^b N_y dN_y + 2\int_a^b (L_T^y - L_S^y)dy.$$

Let $\psi(t,y) = N_y I_K(t,y)$. Then ψ is identifiable, and $\int_a^b N_y dN_y = \iint_K \psi dL$. Since $J_T = N_b$, this proves (3.3). *Q.E.D.*

LEMMA 3.2. Let $\phi(t,x)$ be identifiable and let $h(x)$ be predictable relative to the fields $(\mathcal{E}_x)$. Then

(i) ϕh is identifiable;

(ii) if $\phi \in \mathcal{H}$ and h is bounded, $\phi h \in \mathcal{H}$;

(iii) if, furthermore, $M_x = \iint_{R_x} \phi dL$ then

$$\int_{-\infty}^{x} h(y)dM_y = \iint_{R_x} h(y)\phi(t,y)dL_t^y, \tag{3.4}$$

where

$$R_x = (-\infty,x] \times [0,\infty).$$

PROOF. This is clear if $h(x) = XI_{(y,z]}(x)$, where X is bounded and $\mathcal{E}_y$-measurable. Since processes of this type generate the $\mathcal{E}_x$-predictable processes, the general case follows from a passage to the limit on both sides of (3.4). *Q.E.D.*

LEMMA 3.3. Let $\phi_j \in \mathcal{H}$, $j = 1,2,\dots,n$ and let f be a bounded $C^{(2)}$-function of compact support on $\mathbb{R}^{n+1}$. Let $M_j(x) = \iint \phi(t,y)dL_t^y$, $j = 1,2,\dots,n$, and put, for any function g on $\mathbb{R}^{n+1}$, $\hat{g}(x) = g(x,M_1(x),\dots,M_n(x))$. Then

$$\hat{f}(x) = \iint_{R_x} \sum_{i=1}^{n} \widehat{\frac{\partial f}{\partial x_j}}(y)\phi_j(t,y)dL_t^y + \frac{1}{2}\int_{-\infty}^{x} \widehat{\frac{\partial f}{\partial y}}(y)dy \tag{3.5}$$

$$+ \tfrac{1}{2} \int_{-a}^{x} \Big(\sum_{i,j=1}^{n} \widehat{\frac{\partial^2 f}{\partial x_i \partial x_j}}(y) \int_0^\infty \phi_i(t,y)\phi_j(t,y)\partial_t L_t^y \Big) dy.$$

PROOF. f has compact support, so $\hat{f}(y) = 0$ for large negative y. By Ito's lemma,

$$\hat{f}(x) = \sum_i \int_{-\infty}^{x} \widehat{\frac{\partial f}{\partial x_j}}(y) dM_j(y) + \tfrac{1}{2} \sum_{i,j} \int_{-\infty}^{x} \widehat{\frac{\partial^2 f}{\partial x_i \partial x_j}}(y) d\langle M_i, M_j\rangle y + \int_{-\infty}^{x} \widehat{\frac{\partial f}{\partial y}}(y) dy.$$

Now apply Lemma 3.2 and Theorem 2.5. *Q.E.D.*

Let us mention two special cases which we will need later.

(3.6) $$M_1(x)M_2(x) = \iint_{R_x} (M_1(y)\phi_2(t,y) + M_2(y)\phi_1(t,y)) dL_t^y + \int_{-\infty}^{x} \int_0^\infty \phi_1(t,y)\phi_2(t,y)\partial_t L_t^y dy$$

(3.7) $$g(x)M_1(x) = \iint_{R_x} g(y)\phi_1(y,t) dL_t^y + \int_{\infty}^{x} g'(y) M_1(y) dy.$$

The final result of this section is a disguised form of Fubini's theorem.

PROPOSITION 3.4. Let M be a real number and let $\psi(t,x,y)$ be a process which vanishes for $y < M$ and which, as a function of $(t,x,y;\omega)$, is measurable with respect to the σ-field on $\mathbb{R}_+ \times \mathbb{R} \times \mathbb{R} \times \Omega$ generated by processes of the form $a(t,x)b(y)$, where a is an identifiable process and b is continuous and adapted to the $(\mathcal{E}_x)$. Suppose $E\{\iint_{R_z} \psi^2(t,B_t,y) dt dy\} < \infty$. Then

(3.8) $$\int_{-\infty}^{z} \Big(\iint_{R_z - R_y} \psi(t,x,y) d_{tx} L_t^x \Big) dy = \iint_{R_z} \Big(\int_{-\infty}^{x} \psi(t,x,y) dy \Big) d_{tx} L_t^x.$$

PROOF. First suppose $\psi(t,x,y) = a(t,x)b(y)$, where a and b are bounded and continuous, a is identifiable, and b is adapted to E_y. Suppose a and b vanish if $x < M$ or $y < M$. Fix an integer N and let $M = y_0 < y_1 < \cdots < y_N = z$. Then

$$\iint_{R_z - R_y} \psi(t,x,y)\,dL_t^x = b(y) \iint_{R_z - R_y} a(t,x)\,dL_t^y.$$

We can choose a version of this which is continuous in y. The left-hand side of (3.8) is the limit of the Riemann sums

$$\sum_{n=0}^{N-1} \Big(\iint_{R_z - R_{y_{n+1}}} a(t,x)\,dL_t^x\Big) \big(y_{n+1} - y_n\big) b\big(y_{n+1}\big)$$

$$= \sum_{k=1}^{N-1} \sum_{n=0}^{k-1} \Big(\iint_{R_{y_{k+1}} - R_{y_k}} a(t,x)\,dL_t^x\Big) b\big(y_{n+1}\big)\big(y_{n+1} - y_n\big).$$

Let $J_N(y_k) = \sum_{n=0}^{k-1} b(y_{n+1})(y_{n+1} - y_n)$. The above equals

$$\sum_{k=1}^{N-1} \iint_{R_{y_{k+1}} - R_{y_k}} J_N(y_k) a(t,x)\,dL_t^x.$$

Define $J_N(y) = J_N(y_k)$ if $y_k < y \le y_{k+1}$, and write:

$$= \iint_{R_z} J_N(x) a(t,x)\,dL_t^x.$$

Now let $N \to \infty$ such that the mesh of the partition goes to zero. Then $J_N(x) \to J(x) = \int_{-\infty}^x b(y)\,dy$. Moreover

$$E\Big\{\Big(\iint_{R_z} (J_N(x) - J(x)) a(t,x)\,dL_t^x\Big)\Big\} = 4E\Big\{\int_0^\infty a^2(t,B_t)(J_N(B_t) - J(B_t))^2 dt\Big\}.$$

Now b has compact support, so J and J_N are uniformly bounded. Since a

has compact support, $\int_0^\infty a^2(t,B_t)dt$ is uniformly bounded, so we can conclude that the above integral converges to zero. Moreover, $J(x)a(t,x)$ is clearly identifiable. Thus

$$\iint_{R_z} J_N(x)a(t,x)dL_t^x \to \iint_{R_z} J(x)a(t,x)dL_t^x$$

in L^2, which proves (3.8) in this case.

If ψ satisfies the conditions of the theorem, we can find a sequence of ψ_n which are each finite sums of functions of the form $a(t,x)b(y)$, such that

$$E\{\iint_{R_z}(\psi_n(t,B_t,y) - \psi(t,B_t,y))^2 dtdy\} \to 0. \tag{3.9}$$

Now (3.8) holds for each ψ_n. Note that for a.e. y,

$$\int_0^\infty (\psi_n(t,B_t,y) - \psi(t,B_t,y))^2 dt \to 0 \text{ in } L^2,$$

so that, for a.e. y,

$$\iint_{R_z-R_y} \psi_n(t,x,y)d_{tx}L_t^x \to \iint_{R_z-R_y} \psi(t,x,y)d_{tx}L_t^x \text{ in } L^2.$$

We can then extract a subsequence which converges both a.e. and in L^2, simultaneously for a.e. y. Since ψ_n and ψ both vanish for $y < M$, the integral over y is only over a finite interval, and the above implies that the family $\iint_{R_z-R_y} \psi(t,x,y)d_{tx}L_t^x$ is a.s. uniformly integrable as a function of y. Thus the left-hand side of (3.8) converges in L^2.

On the right-hand side

$$E\{(\iint_{R_z}\int_{-\infty}^{x}(\psi_n(t,x,y) - \psi(t,x,y))dy\,dL_t^x)^2\}$$

$$= 4E\{\int_0^\infty(\int_{-\infty}^{B_t}(\psi_n(t,B_t,y) - \psi(t,B_t,y))dy\,I_{\{B_t \le z\}})^2 dt\}.$$

Both ψ_n and ψ vanish for $y < M$, so by Schwartz' inequality, this is

$$\le 4(z-M)\ E\{\int_0^\infty\int_{-\infty}^{z}(\psi(t,B_t,y) - \psi_n(t,B_t,y))^2 dy\ dt\}$$

which converges to zero by (3.9). Thus the right-hand side of (3.8) also converges in L^2. *Q.E.D.*

§4. Integrals with respect to $\partial_y L_t^y$

Let us consider L_t^y as a function of y for fixed t. Let $G_{ty} = \sigma\{B_{\rho(s,y)}: s \le H_t^y\}$. This is the field generated by the excursions below y, up to time t. Notice that $B_t \wedge y$ and $\{L_t^x: x \le y\}$ are measurable with respect to G_{ty}. Perkins [2] has shown that for each fixed t, $\{L_t^y, G_{ty}, y > -\infty\}$ is a semi-martingale, and has given explicit formulas for the process of bounded variation $\{V_t(y), y > -\infty\}$ in the decomposition

(4.1) $$L_t^y = M_t(y) + V_t(y), \quad y > -\infty$$

of L_t^y into a local martingale $M_t(y)$ plus a process of bounded variation $V_t(y)$.

Let us compare the fields G_{ty} and $\mathcal{E}_y$. Let $\hat{G}_{ty} = \sigma\{B_{\rho(s,y)}, s \geq H_t^y\}$ and

$$\hat{\mathcal{E}}_{ty} = \mathcal{E}_y \vee \sigma\{L_t^x, x \leq y\}.$$

Perkins has shown that $G_{ty} \vee \hat{G}_{ty} = \hat{\mathcal{E}}_{ty}$. Now G_{ty} and $\hat{G}_{ty}$ are conditionally independent given $B_t \wedge y$, which follows from the strong Markov property of B. It follows that $\{L_t^y, \hat{\mathcal{E}}_{ty}, y > -\infty\}$ is also a semi-martingale with the same decomposition (4.1).

We will not need the exact form of $V_t(y)$, but we will need the following [2].

THEOREM 4.1. (Perkins) (i) $V_{ty} = \int_{-\infty}^{y} v(t,x)dx$, where for each t and $0 < p < \infty$, $\sup_y E\{|v(t,y)|^p\} < \infty$; (ii) $\langle M_t \rangle_y = 4 \int_{-\infty}^{y} L_t^x dx$.

Since L_t^y is a semi-martingale, we can define stochastic integrals with respect to it. We can integrate $\hat{\mathcal{E}}_{ty}$-predictable functions. Here is one source of such functions.

PROPOSITION 4.2. Let S be an $\mathcal{E}_x$-identifiable r.v., and let $\phi(t,y)$ be an identifiable process. Then for any $t_0 > 0$

(i) $\{S < t_0\}$, $\{s = t_0\}$ and $\{S > t_0\}$ are in $\hat{\mathcal{E}}_{t_0 x}$;

(ii) $\{\phi(t_0,y), y > -\infty\}$ is predictable relative to the fields $(\hat{\mathcal{E}}_{t_0 y}, y > -\infty)$.

PROOF. H_t^x is an increasing function of t, and, by Lemma A2 of the appendix, H_t^x is strictly increasing both at S^- and at S^+ on the set $\{S > 0\}$. Consequently, on the set $\{S > 0\}$, $S < t_0$ iff $H_S^x < t_0$ and $S = t_0$ iff $H_S^x = t_0$. Thus

$$\{S < t_0\} = \{S = 0\} \cup \{H_S^x < H_{t_0}^x\}, \quad \text{and}$$

$$\{S = t_0\} = \{H_S^x = H_{t_0}^x\}.$$

But $\{S = 0\} \in \mathcal{E}_x$, H_S^x is $\mathcal{E}_x$-measurable by Proposition 2.1, and $H_{t_0}^x = \int_{-\infty}^x L_{t_0}^y\, dy$, which is $\hat{\mathcal{E}}_{t_0 x}$-measurable. This implies (i).

To prove (ii), note that if ϕ is of the form

$$\phi(t,y) = X\, I_{\{S < t \le T;\ a < y \le b\}}$$

where X is $\mathcal{E}_a$-measurable and S and T are $\mathcal{E}_a$-identifiable, then $\phi(t_0,y) \equiv 0$ for $y \le a$, and for $y > a$, $\{S < t_0 \le T\} \in \hat{\mathcal{E}}_{t_0 a} \subset \hat{\mathcal{E}}_{t_0 y}$ by part (i), so that $\{\phi(t,y),\ y > -\infty\}$ is adapted to $(\hat{\mathcal{E}}_{t_0 y})$ and hence, being left-continuous, is $\hat{\mathcal{E}}_{t_0 y}$-predictable. The class of identifiable processes is generated by processes of this form, and (ii) follows.

Q.E.D.

Let ϕ be identifiable and define

$$\Phi(t,z) = \int_{-\infty}^{z} \phi(t,y)\, \partial_y L_t^y\,; \tag{4.2}$$

$$\Phi_m(t,z) = \int_{-\infty}^{z} \phi(t,y)\, \partial_y M_t(y); \tag{4.3}$$

$$\Phi_v(t,z) = \int_{-\infty}^{z} \phi(t,y) v(t,y)\, dy, \tag{4.4}$$

where M and V are the processes of (4.1), and $v(t,y)$ is the derivative of V. If Φ_m and Φ_v exist, so does Φ, and $\Phi = \Phi_m + \Phi_v$. Since $\phi(t,\cdot)$ is $\hat{\mathcal{E}}_{ty}$-predictable, Φ_m exists if

$$\int_{-\infty}^{z} \phi^2(t,y)\, d\langle M_t\rangle_y = 4\int_{-\infty}^{z} \phi^2(t,y) L_t^y\, dy < \infty \quad \text{a.s.}$$

Since $y \to L_t^y$ is bounded and of compact support (it vanishes unless $\min_{s \le t} B_s \le y \le \max_{s \le t} B_s$) a sufficient condition for the existence of Φ_m is that $\int \phi^2(t,y)dy < \infty$.

By Theorem 4.1, there exists a function $C(t,p)$ such that $E\{|v(t,y)|^p\} \le C(t,p)$ for all $p > 1$. If $\frac{1}{p} + \frac{1}{q} = 1$, Hölder's inequality gives

$$\int_{-\infty}^{z} |\phi(t,y)v(t,y)|\,dy \le \Big(\int_{m(t)}^{z} |\phi(t,y)|^q dy\Big)^{1/q}\Big(\int_{m(t)}^{z} |v(t,y)|^p dy\Big)^{1/p}$$

where $m(t) = \min_{s \le t} B_s$. Thus

$$E\{\int_{-\infty}^{z} |\phi(t,y)v(t,y)|\,dy\} \le E\{\int_{m(t)}^{z} |\phi(t,y)|^q dy\}E\{\int_{m(t)}^{z} |v(t,y)|^p dy\}^{1/p}$$

$$\le E\{|z - m(t)|\}E\{\int_{-\infty}^{z}|\phi(t,y)|^{2q}dy\}^{\frac{1}{2}q}\ C(2p,t).$$

It follows that a sufficient condition for the existence of Φ_v is that $E\{\int_{-\infty}^{z} |\phi(t,y)|^{1+\varepsilon}dy\} < \infty$ for some $\varepsilon > 0$.

Let us combine these two to get a single sufficient — but far from necessary — condition for the existence of Φ.

PROPOSITION 4.3. Let ϕ be identifiable and let $t > 0$. A sufficient condition for the existence of $\Phi(t,z)$ for all z is that for each $a < b$, $\int_a^b \phi^2(t,y)dy < \infty$ a.s.

PROOF. Suppose first that $\phi(t,y) = 0$ for all $y < a$, for some fixed a. Then $\Phi_m(t,z)$ is defined if $\int_a^z \phi^2(t,y)dy < \infty$.

Now let $Z_N = \inf\{x: \int_a^x \phi^2(t,y)dy > N\}$. Then $\int_{-\infty}^{z} \phi(t,y)v(t,y)I_{\{y \le Z_n\}}dy$ exists a.s. by our above remarks. Let $N \to \infty$ and notice that $Z_N \to \infty$ a.s. and that therefore $\int_{-\infty}^{z} \phi(t,y)v(t,y)dy$

exists, since it equals the above integral on the set $\{Z_n \geq z\}$.

To remove the restriction that $\phi(t,y)$ vanish for $y < a$, we just remark that $L_t^y \equiv 0$ on $y < m(t) = \min_{s \leq t} B_s$. Thus for each ω,

$$\int_{-\infty}^{z} \phi \partial_y L_t^y = \int_{a}^{z} \phi \partial_y L_t^y$$

on the set $\{m(t) > a\}$. *Q.E.D.*

We will need some L^p-estimates for integrals involving the functions ϕ and Φ in the next section. Let us define a global norm $\|\phi\|_4$ by

$$(4.5)\quad \|\phi\|_4 = E\{\int_0^\infty \phi(t,B_t)^4 dt\}^{\frac{1}{4}} + E\{\int_{-\infty}^{\infty}\int_0^\infty \phi(t,y)^4 dtdy\}^{\frac{1}{4}} + E\{\left(\int_{-\infty}^{\infty}\int_0^\infty \phi(t,y)^2 dtdy\right)^2\}^{\frac{1}{4}}.$$

PROPOSITION 4.4. Let ϕ be identifiable and suppose there exist real a, b and N such that $\phi(t,x) = 0$ unless $a \leq x \leq b$ and $t \leq T_{-N}$. Suppose $\|\phi\|_4 < \infty$. Then for a.e. t, $\Phi(t,z)$ exists for all z, is a.s. continuous in z, and there exist constants A, B and C (which may depend on a, b and N) such that

(i) $$E\{\sup_{y \leq b} |\Phi(t,y)|^2\}^2 \leq A\, E\{\int_{-\infty}^{b} \phi(t,y)^4 dy\};$$

(ii) $$E\{\int_{-\infty}^{z}\int_0^\infty |\Phi(t,y)|\partial_t L_t^y dy\} \leq B\,\|\phi\|_4^2\ ;$$

(iii) $$E\{\int_{-\infty}^{z}\int_0^\infty \left(\int_t^\infty \phi(s,y)\partial_s L_s^y\right)^2 \partial_t L_t^y dy\} \leq C\,\|\phi\|_4^2\ .$$

PROOF. If $\|\phi\|_y < \infty$, $\int_a^b \phi^2(t,y)dy < \infty$ a.s. for a.e. t by Fubini's theorem so that for a.e. t, $\Phi(t,z)$ exists a.s. for all z. Fix t. By Doob's inequality,

$$E\{\sup_{y \leq b} \Phi_m(t,y)\} = E\{\sup_{y \leq b}\left(\int_{-\infty}^{b} \phi(t,x)\partial_x M_t(x)\right)^2\} \leq 4E\{\int_{-\infty}^{b} \phi^2 d\langle M\rangle\}$$

$$= 16\ E\{\int_{-\infty}^{b} \phi^2(t,y)L_t^y dy\}.$$

Let $T = T_{-N}$. Then $\phi = 0$ if $t > T$, so $L_t^y \le L_T^y$:

$$\le 16\ E\{\int_a^b \phi^2(t,y)L_T^y dy\}$$

$$\le 16\ E\{\int_a^b (L_T^y)^2 dy\}^{\frac{1}{2}}\ E\{\int_a^b \phi^4(t,y)dy\}^{\frac{1}{2}}$$

$$= 16 \int_a^b E\{(L_T^y)^2\}dy\ E\{\int_a^b \phi^4(t,y)dy\}^{\frac{1}{2}}.$$

Now, given that B_t hits y before it hits $-N$, L_T^y is exponential for all y, and it is not hard to see that its parameter is bounded away from zero in $[a,b]$, and hence that the next-to-last term above is finite. Let $A_1 = 16\int_a^b E\{(L_T^y)^2\}dy$. Next,

$$\sup_{y \le b} \Phi_v(t,y) \le \int_a^b |\phi(t,y)v(t,y)|dy.$$

Thus, by Schwartz' inequality

$$E\{\sup_{y \le b} \Phi_v(t,y)^2\} \le E\{\int_a^b \phi^2(t,y)dy\}^{\frac{1}{2}}\ E\{\int_a^b v^2(t,y)dy\}^{\frac{1}{2}}$$

$$\le (b-a)^{\frac{1}{2}}\ E\{\int_a^b v^2(t,y)dy\} E\{\int_a^b \phi^4(t,y)dy\}^{\frac{1}{4}}$$

$$= A_2\ E\{\int_a^b \phi^4(t,y)dy\}^{\frac{1}{4}}$$

where we have used the fact that $v(t,y)$ is uniformly L^2-bounded. Now (i) follows with $A = 2A_1 + 2A_2$.

Moving to (ii), we write

$$(4.6)\qquad \int_{-\infty}^{z}\int_{0}^{\infty}|\Phi(t,y)|\,\partial_t L_t^y dy = \int_0^\infty |\Phi(t,B_t)| I_{\{B_t \le z\}} dt$$

$$\le \int_0^T \sup_{y\le z}|\Phi(t,y)| I_{\{B_t \le z\}} dt$$

$$\le \Big(\int_0^\infty \sup_{y\le z}|\Phi(t,y)|^2 dt\Big)^{\frac12}\big(H_T^z\big)^{\frac12}.$$

Thus, taking the expectation of both sides and using (i):

$$(4.7)\qquad E\{\int_{-\infty}^{z}\int_{0}^{\infty}|\Phi(t,y)|\,\partial_t L_t^y dy\} \le AE\{H_T^z\}^{\frac12}\, E\{\int_0^\infty\int_{-\infty}^{z}\phi(t,y)^4 dy dt\}^{\frac12}.$$

But $E\{H_T^z\} < \infty$, and the last expectation is bounded by $\|\phi\|_4^2$.

Going on to (iii), we descend once more into the Schwartz pit:

$$(4.8)\qquad \Big(\int_t^\infty \phi(s,y)\partial_s L_s^y\Big)^2 \le \big(L_T^y - L_t^y\big)\int_t^T \phi(s,y)^2 \partial_s L_s^y .$$

Now $\int (L_T^y - L_t^y)\partial_t L_t^y = L_T^y L_t^y - \frac12 (L_t^y)^2$. Integrating by parts

$$\int_0^T \big(L_T^y - L_t^y\big)\int_t^T \phi^2(s,y)\partial_s L_s^y \partial_t L_t^y = \int_0^T \big(L_T^y L_t^y - \tfrac12 (L_t^y)^2\big)\phi^2(t,y)\partial_t L_t^y$$

$$\le \Big(\int_0^T (L_T^y L_t^y - \tfrac12 (L_t^y)^2)\partial_t L_t^y\Big)^{\frac12}\Big(\int_0^\infty \phi(t,y)^4 \partial_t L_t^y\Big)^{\frac12}$$

$$= \Big(\frac{2}{15}\Big)^{\frac12}\big(L_T^y\big)^{5/2}\Big(\int_0^\infty \phi(t,y)^4 \partial_t L_t^y\Big)^{\frac12}.$$

Integrate over y from a to z and take expectations of both sides. Then the left-hand side of (iii) is

$$\le \Big(\frac{2}{15}\Big)^{\frac12} E\{\int_0^T \big(L_t^{B_t}\big)^4 I_{\{a\le B_t \le z\}} dt\}^{\frac12} E\{\int_0^\infty \phi^4(t,B_t) dt\}^{\frac12}.$$

The first expectation is finite (it is bounded by $E\{(H_T^z)^2\}^{\frac12} E\{\sup_{a<y<z} (L_T^y)^8\}$) while the second expectation is bounded by $\|\phi\|_4^2$. *Q.E.D.*

§5. First-Order Green's Formulas

We will establish several formulas which relate line and area integrals. We call them Green's formulas, although the analogy with the classical theorems is perhaps a bit strained. They are rather complex; we will deal with the case of simple functions first. We let R_z denote the quarter-space $R_z = \{(t,x): t \geq 0,\ x \leq z\}$.

Let $\phi \in \mathcal{H}$ and define

$$\Phi(t,x) = \int_{-\infty}^{x} \phi(t,y)\partial_y L_t^y \quad \text{and} \quad \Psi(t,x) = \int_{-\infty}^{x} \phi(t,y)dy.$$

THEOREM 5.1. Suppose ϕ is a simple function and f is a bounded Borel function on $\mathbb{R}$. Then

$$(5.1)\qquad \int_0^\infty (\Phi(t,z) - \Phi(t,B_t)f(B_t)\partial_t H_t^z = \iint_{R_z}\Big(\int_{-\infty}^{x}\int_t^\infty \phi(s,x)f(y)\partial_s L_s^y dy\Big)dL_t^x$$

$$(5.2)\qquad \int_0^\infty \Phi(t,z)\partial_t L_t^z - \iint_{R_z}\Big(\Phi(t,y) + \int_t^\infty \phi(s,y)\partial_s L_s^y\Big)dL_t^y + \int_\infty^z\int_0^\infty \phi(t,y)\partial_t L_t^y dy$$

$$(5.3)\qquad \int_0^\infty \Psi(t,z)\partial_t L_t^z = \iint_{R_z} \Psi(t,y)dL_t^y + \int_{-\infty}^{z}\int_0^\infty \phi(t,y)\partial_t L_t^y dy.$$

PROOF. Suppose $\phi(t,y) = I_{(S,T]}(t)\ I_{(a,b]}(y)$ where S and T are a proper pair of $\mathcal{E}_a$-identifiable times. Then $\phi(t,y) = (L_t^{y\wedge b} - L_t^{y\wedge a}) \cdot I_{(S,T]}(t)$, if $y > b$, so both sides of (5.1) vanish if $z < a$. Rewrite the left-hand side in the form

$$(5.4)\qquad \int_{-\infty}^{z}\int_0^\infty (\Phi(t,z) - \Phi(t,x))\partial_t L_t^x\ f(x)dx.$$

The integrand vanishes if $b < x \leq z$. The same is true of the right-hand side, so that both sides are constant functions of z for $z > b$. Thus we need only prove (5.1) in case $a < z \leq b$.

Let $J_t = L_t^z - L_t^{y\vee a} - L_S^z + L_S^{y\vee a}$ for $t > S$. Then (5.4) is

$$= \int_{-\infty}^{z} (L_S^z - L_S^{y\vee a})(L_T^y - L_S^y)f(y)dy + \int_{-\infty}^{z}\int_S^T J_t \partial_t L_t^y dy.$$

Rewrite the first integral and apply (3.2) to the second:

$$= \int_{-\infty}^{z}\Big(\int_0^S \int_{y\vee a}^{z} \big(L_T^y - L_S^y\big) d_{tx} L_t^x\Big) f(y)dy + \int_{-\infty}^{z}\Big(\int_S^T \int_{y\vee a}^{z} \big(L_T^y - L_t^y\big) d_{tx} L_t^x\Big) f(y)dy$$

$$= \int_{-\infty}^{z}\int_0^T \int_{y\vee a}^{z} \big(L_T^y - L_{S\vee t}^y\big) f(y) d_{tx} L_t^x dy.$$

If $a < x \le z$ and $t \le T$, $L_T^y - L_{S\vee t}^y = \int_t^\infty \phi(s,x)\partial_s L_s^y$. This last integral vanishes if $x \le a$, so the above is

$$= \int_{-\infty}^{z}\int_0^\infty \int_y^z f(y) \int_t^\infty \phi(s,x)\partial_s L_s^y d_{tx} L_t^x dy.$$

By Proposition 3.4, this is

$$= \iint_{R_z}\Big(\int_{-\infty}^{x} f(y) \int_t^\infty \phi(s,x)\, \partial_s L_s^y dy\Big) d_{tx} L_t^x.$$

Moving on to (5.2), note that both sides vanish if $z < a$. Suppose $a < z \le b$ and write $\partial_t L_t^z = \partial_t(L_t^z - L_t^a) + \partial_t L_t^a$. The left-hand side of (5.2) is

(5.5) $$= \int_S^T (L_t^z - L_t^a)\partial_t(L_t^z - L_t^a) + \int_S^T (L_t^z - L_t^a)\partial_t L_t^a.$$

The first Stieltjes integral is, by direct evaluation,

$$\tfrac{1}{2}(L_T^z - L_T^a)^2 - \tfrac{1}{2}(L_S^z - L_S^a)^2.$$

Use Ito's formula to write this as an integral over y:

$$= \int_a^z (L_T^y - L_T^a)\partial_y L_S^y - \int_a^z (L_S^y - L_S^a)\partial_y L_S^y + 2\int_a^z (L_T^y - L_S^y)dy.$$

Now let $K = (S,T] \times (a,b]$, $K_0 = (0,S] \times (a,b]$ and $K_1 = (S,T] \times (-\infty,a]$. Use Lemma 3.2 to convert these to area integrals:

$$= \iint_{R_z} [I_{K_0}(L_T^y - L_S^y) + I_K(L_T^y - L_t^y) + I_K(L_t^y - L_T^a)$$
$$- I_{K_0}(L_T^a - L_S^a)]dL_t^y + 2\int_a^z (L_T^y - L_S^y)dy.$$

Now
$$\int_t^\infty \phi(s,y)\partial_s L_s^y = \begin{cases} L_T^y - L_S^y & \text{if } (y,t) \in K_0 \\ L_T^y - L_t^y & \text{if } (y,t) \in K . \\ 0 & \text{otherwise} \end{cases}$$

The above expression becomes

$$\iint_{R_z} [\int_t^\infty \phi(s,y)\partial_s L_s^y + I_K(L_t^y - L_T^a) - I_{K_0}(L_T^a - L_S^a)]dL_t^y + 2\int_{-\infty}^z \int_0^\infty \phi(s,y)\partial_s L_s^y dy.$$

Now by Proposition 3.1 the final integral in (5.5) equals

$$\iint_{R_z} I_K(L_T^a - L_t^a)dL_t^y + (L_S^z - L_S^a)(L_T^a - L_S^a).$$

Add these two expressions. After cancellation, we see that (5.5) is

$$= \iint_{R_z} (\int_t^\infty \phi(s,y)\partial_s L_s^y + I_K(L_t^y - L_t^a))dL_t^y + 2\int_{-\infty}^z \int_0^\infty \phi(s,y)\partial_s L_s^y dy.$$

But $\Phi(t,y) = I_K(L_t^y - L_t^a)$ for $a \le y \le b$, and so (5.2) follows immediately for $z \le b$.

Finally, if $z > b$, $\Phi(t,z) = \Phi(t,b)$, so that

$$\int_0^\infty \Phi(t,z)\partial_t L_t^z = \int_0^\infty \Phi(t,b)\partial_t L_t^b + \iint_{R_z - R_b} \Phi(t,b)dL_t^y .$$

Apply (5.2) (with z = b) to the first integral on the right, and note that ϕ vanishes on $R_z - R_b$ to see that (5.2) holds for $z > b$ as well.

Now let us look at (5.3). Both sides vanish if $z < a$. If $z \geq a$, $\Psi(t,y) = (y \wedge b - a)I_{(S,T]}(y)$. Thus the left-hand side equals

$$(z \wedge b - a)(L_T^z - L_S^z) = \int_a^z (y \wedge b - a)\partial_y (L_T^y - L_S^y) + \int_a^z (L_T^{y\wedge b} - L_S^{y\wedge b})dy$$

by Ito's lemma. Apply Lemma 3.2 and put in Ψ and ϕ:

$$= \iint_{R_z - R_a} \Psi(t,y)dL_t^y + \int_a^z \Big(\int_0^\infty \phi(t,y)\partial_t L_t^y\Big) dy. \qquad \textit{Q.E.D.}$$

We now extend Theorem 5.1 in a form which looks somewhat more like the classical Green's theorem. We will state it for the infinite strip $R_b - R_a = [0,\infty) \times (a,b]$.

Let $a < b$, let $\phi \in \mathcal{H}$, let $\Phi(t,x)$ and $\Psi(t,x)$ be adapted to $(\mathcal{E}_x)$ and satisfy, for $x > a$,

(5.6) $$\Phi(t,x) = \Phi(t,a) + \int_a^x \phi(t,y)\partial_y L_t^y$$

(5.7) $$\Psi(t,x) = \Psi(t,a) + \int_a^x \phi(t,y)dy.$$

THEOREM 5.2. Let $z < a \leq b$ and let $\phi \in \mathcal{H}_4$ be such that $\phi(t,x) = 0$ for all x if $t > T_z$. Let Φ and Ψ be identifiable processes satisfying (5.6) and (5.7). We suppose that the initial values of Φ and Ψ satisfy

$$\int |\Phi(t,a)|\partial_t L_t^a < \infty, \quad \int |\Psi(t,a)|\partial_t L_t^a < \infty,$$

$$\int |\Phi(t,a)|^2 \partial_t H_t^b < \infty, \quad \int |\Psi(t,a)|^2 \partial_t H_t^b < \infty.$$

Let f be a bounded Borel function. Then,

$$(5.8)\quad \int_a^b \int_0^\infty (\Phi(t,b) - \Phi(t,y))\partial_t L_t^y f(y)dy = \iint_{R_b - R_a} \left(\int_a^x \int_t^\infty \phi(s,x)f(y)\partial_s L_s^y dy\right) dL_t^x$$

$$(5.9)\quad \int_0^\infty \Phi(t,b)\partial_t L_t^b - \int_0^\infty \Phi(t,a)\partial_t L_t^a = \iint_{R_b - R_a} (\Phi(t,y) + \int_t^\infty \phi(s,y)\partial_s L_s^y) dL_t^y$$

$$+ 2\int_a^b \int_0^\infty \phi(t,y)\partial_t L_t^y dy$$

$$(5.10)\quad \int_0^\infty \Psi(t,b)\partial_t L_t^b - \int_0^\infty \Psi(t,a)\partial_t L_t^a = \iint_{R_b - R_a} \Psi(t,y)dL_t^y + \int_a^b \int_0^\infty \phi(t,y)\partial_t L_t^y dy.$$

NOTE. It is (5.9) and (5.10) which resemble Green's theorem, rather than (5.8). The left-hand sides of these can be thought of as line integrals around the boundary. The reason for the profusion of terms on the right-hand sides is simply that dL is not a product measure.

PROOF. Since $\phi \in H_4 \subset H$ and since H is generated by simple functions, it follows by standard arguments that there exist simple ϕ_n such that $\|\phi - \phi_n\|_4 \to 0$. Define

$$\Phi_n(t,x) = \Phi(t,a) + \int_a^x \phi_n(t,y)\partial_y L_t^y.$$

Note that $\Phi(t,a)$ cancels out in (5.8), so we may as well suppose it vanishes. Then (5.8) holds for each Φ_n: just set $f(y) = 0$ for $y < a$ in (5.1). Now put $\|f\| = \sup|f(x)|$ and consider the left-hand side of (5.8). By Proposition 4.4(i),

$$E\{\int_a^b \int_0^\infty |\Phi(t,y) - \Phi_n(t,y)|\partial_t L_t^y f(y)dy\}$$

$$\leq \|f\| E\{\int_a^b \int_0^\infty |\Phi(t,y) - \Phi_n(t,y)|\partial_t L_t^y dy\} \leq B\|f\| \|\phi - \phi_n\|_4^2,$$

which converges to zero. Moreover

$$|\Phi(t,b) - \Phi_n(t,b)| \le \sup_{y \le b} |\Phi(t,y) - \Phi_n(t,y)|,$$

so, from (4.6) and (4.7)

$$E\{\int_a^b |\Phi(t,b) - \Phi_n(t,b)| \partial_t L_t^y f(y)dy\} \le B \|\phi - \phi_n\|_4^2 .$$

These two together show that

$$\int_a^b \int_0^\infty (\Phi_n(t,z) - \Phi_n(t,y))\partial_t L_t^y f(y)dy$$

converges in L^1 and hence in measure to the corresponding integral for Φ.

On the right-hand side, put $\chi_n(t,x) = \phi_n(t,x) - \phi(t,x)$. Then

$$E\{\iint_{R_b - R_a} (\int_t^\infty \chi_n(s,x) f(B_s) I_{\{B_s \le x\}} ds) dL_t^x\}$$

$$= E\{\int_0^\infty (\int_t^\infty \chi_n(s,B_t) f(B_s) I_{\{a \le B_s \le B_t \le z\}} ds)^2 dt\}$$

$$\le \|f\| E\{\int_0^\infty (\int_0^T \chi_n(s,B_t) ds I_{\{a \le B_s \le B_t \le z\}})^2 dt\}.$$

But

$$(\int_0^T \chi_n(s,B_t) I_{\{a \le B_s \le B_t \le z\}} ds)^2 \le (\int_0^T I_{\{a \le B_s \le B_t \le z\}} ds)(\int_0^T \chi_n^2(s,B_t) ds).$$

Let $V_n(y) = E\{\int_0^T \chi_n^2(s,y)ds\}$. The above expectation is

$$\le \|f\| E\{(\int_0^T I_{\{a \le B_s \le b\}} ds)(\int_0^T V_n(B_t)dt)\}$$

$$= \|f\| E\{\int_0^T I_{\{B_s \in (a,b)\}} ds \int_a^b V_n(y) L_T^y dy\}$$

$$\leq \|f\| \, E\{\int_0^T I_{\{a \leq B_s \leq b\}} ds (\sup_{a \leq y \leq b} L_T^y) \int_a^b V_n(y)dy\}.$$

But $\int_a^b V_n(y)dy \leq \int_{-\infty}^{\infty}\int_0^{\infty} \chi_n^2(s,y)dsdy$; the expected square of this integral is dominated by $\|\phi_n - \phi\|_4^2$. Thus the above is

$$\leq \|f\| \, E\{(\int_0^T I_{\{a \leq B_s \leq z\}} ds)^2 \sup_{a \leq y \leq b} (L_T^y)^2\} \, \|\phi_n - \phi\|_4^2 .$$

This tends to zero. Thus the right-hand side of (5.8) converges in L^2 to the proper value. This proves (5.8).

Now consider (5.9). By linearity, it is enough to verify this separately in the two special cases $\phi \equiv 0$ and $\Phi(\cdot,a) \equiv 0$. If $\phi \equiv 0$, (5.9) becomes

$$\text{(5.11)} \qquad \int_0^{\infty} \Phi(t,a)\partial_t(L_t^b - L_t^a) = \iint_{R_b - R_a} \Phi(t,a)dL_t^y,$$

which follows by Corollary 2.12. (This applies since $\int|\Phi(t,a)|\partial_t L_t^a$ is finite, and the fact that $\int\Phi(t,a)^2\partial_t H_t^b < \infty$ implies that $\hat{\Phi} = \Phi I_{\{y \in (a,b)\}}$ is in $H_{\ell oc}$.)

Next, suppose $\Phi(\cdot,a) \equiv 0$. Apply Proposition 4.4(ii) to the left-hand side of (5.9).

$$E\{\int_0^{\infty} |\Phi(t,z) - \Phi_n(t,z)| \, \partial_t L_t^z\} \leq B \, \|\phi - \phi_n\|_4^2$$

which goes to zero. Thus the left-hand side of (5.9) converges in L^1, hence in measure. Each of the three integrals on the right-hand side converges too. The first converges since

$$E\{(\iint_{R_b} (\Phi(t,y) - \Phi_n(t,y))dL_t^y)^2\} = E\{\int_0^{\infty} (\Phi(t,B_t) - \Phi_n(t,B_t))^2 dt\}$$

$$\leq E\{\int_0^\infty \sup_{y \leq b}(\Phi(t,y) - \Phi_n(t,y))^2 dt\}$$

$$\leq A\, E\{\int_a^b \int_0^\infty (\phi(t,y) - \phi_n(t,y))^4 dt dy\}$$

$$\leq A\, \|\phi - \phi_n\|_4^4$$

by Proposition 4.4(i). The second converges since

$$E\{(\iint_{R_b} (\int_t^\infty (\phi(s,y) - \phi_n(s,y))\partial_s L_s^y) dL_t^y)^2\}$$

$$= E\{\int_a^b \int_0^\infty (\int_t^\infty (\phi(s,y)) - \phi_n(s,y)\partial_s L_s^y)^2 \partial_t L_t^y dy\}$$

$$\leq C\, \|\phi\|_4^2$$

by Proposition 4.4(iii). Finally

$$\int_a^b \int_0^\infty |\phi(t,y) - \phi_n(t,y)|\partial_t L_t^y dy = \int_0^T |\phi(t,B_t) - \phi_n(t,B_t)| I_{\{B_t \leq z\}} dt$$

and this converges to zero in measure since $\int_0^T (\phi(t,B_t) - \phi_n(t,B_t))^4 dt$ does (recall that ϕ has compact support in t).

Moving to (5.10), notice that the case $\phi \equiv 0$ reduces exactly to that of (5.9), which we have just treated. Thus suppose $\Psi(\cdot,a) \equiv 0$. As above, approximate ϕ by simple ϕ_n for which $\|\phi - \phi_n\|_4 \to 0$. Then (5.10) holds for each ϕ_n. Let $n \to \infty$. We have just seen that the final integral in (5.10) converges, while the next-to-last converges since

$$E\{(\iint_{R_b - R_a} (\Psi_n(t,y) - \Psi(t,y)) dL_t^y)^2\} = E\{\int_0^\infty (\Psi_n(t,B_t) - \Psi(t,B_t))^2 I_{\{B_t \in (a,b]\}} dt\}$$

$$\leq E\{\int_0^\infty (\int_a^b |\phi_n(t,y) - \phi(t,y)| dy)^2 dt\}$$

$$\leq (b-a)E\{\int_0^\infty \int_a^b (\phi_n(t,y) - \phi(t,y))^2 dydt\}$$

which converges to zero since the expected square of the double integral is bounded by $||\phi_n - \phi||_4^4$.

Finally, integrate the left-hand side of (5.10) with respect to b from α to β.

$$\int_\alpha^\beta \int_0^T |\Psi(t,b) - \Psi_n(t,b)| \partial_t L_t^b db = \int_0^T |\Psi(t,B_t) - \Psi_n(t,B_t)| I_{\{\alpha \leq B_t \leq \beta\}} dt$$

$$\leq \int_0^T \int_\alpha^\beta |\phi(t,y) - \phi_n(t,y)| dydt$$

which tends to zero in measure since $\iint(\phi(t,y) - \phi_n(t,y))^2 dydt$ converges to zero in L^2. It follows that there exists a subsequence (n_k) such that

$$\int_0^\infty \Psi_{n_k}(t,b)\partial_t L_t^b \to \int_0^\infty \Psi(t,b)\partial_t L_t^b \quad \text{a.s.}$$

for a.e. b. This proves (5.10) for a.e. b. Now fix a < b. For a.e. $c < b+\varepsilon$, (5.10) holds a.s. with b replaced by c. Similarly, if $E\{\int_0^\infty |\Psi(t,b)| \partial_t L_t^b\} < \infty$, then (5.10) will also hold with a replaced by b and b by c. Choose c > b for which both hold simultaneously and subtract. The result is exactly (5.10). It follows that (5.8) holds for any b for which

$$E\{\int_0^\infty |\Psi(t,b)| \partial_t L_t^b\} < \infty,$$

i.e. for which
$$E\{\int_0^\infty \int_a^b \phi(t,y)dy \partial_t L_t^b\} < \infty.$$

Apply Corollary 2.11 noting that $\phi(0,x) = 0$ since $\phi \in H$. The expectation is bounded by

$$\int_a^b E\{\int_0^\infty |\phi(t,y)|\,\partial_t L_t^b\}dy \le E\{\int_a^b \int_0^\infty |\phi(t,y)|\,\partial_t L_t^y dy\}$$

$$= E\{\int_0^\infty |\phi(t,B_t)|\, I_{\{a \le B_t \le b\}} dt\}.$$

By Hölder's inequality, this is

$$\le \|\phi\|_4 \, E\{\int_0^T I_{\{a \le B_t \le b\}} dt\}^{3/4},$$

which is finite. This finishes the proof.

COROLLARY 5.3. Let $\phi \in H_4$, $a < b$, and let Φ and Ψ be defined as in (5.6) and (5.7) and satisfy the conditions of Theorem 5.2. Let f be a bounded Borel function on $\mathbb{R}$ which vanishes on $(-\infty,a]$. Then for each $b > a$

$$(5.12)\qquad \int_0^\infty \Phi(t,b)f(B_t)\partial_t H_t^b = \iint_{R_b - R_a} \Big(\int_a^x \int_t^\infty \phi(s,x)f(y)\partial_s L_s^y dy\Big) dL_t^x$$

$$+ \int_a^b \int_0^\infty \Phi(t,y)f(y)\partial_t L_t^y dy$$

$$(5.13)\qquad \int_0^\infty \Psi(t,b)f(B_t)\partial_t H_t^b = \int_a^b \int_0^\infty \Phi(t,y)f(y)\partial_t L_t^y dy$$

$$+ \int_a^b \int_a^y \int_0^\infty \phi(t,y)f(z)\partial_t L_t^z dz dy.$$

PROOF. $\int_0^t \Phi(t,b)f(B_t)\partial_t H_t^b = \int_a^b \int_0^\infty \Phi(t,b)f(y)\partial_t L_t^y dy$

$$= \int_a^b \int_0^\infty \Phi(t,y)f(y)\partial_t L_t^y dy + \int_a^b \int_0^\infty (\Phi(t,b) - \Phi(t,y))f(y)\partial_t L_t^y dy.$$

Apply (5.8) to the last integral. It equals the first integral on the right-hand side of (5.12).

The left-hand side of (5.13) equals

$$\int_a^b \int_0^\infty \Phi(t,y)f(y)\partial_t L_t^y \, dy + \int_a^b \int_0^\infty (\Psi(t,b) - \Psi(t,y))f(y)\partial_t L_t^y \, dy.$$

The latter integral equals

$$\int_a^b \int_0^\infty \int_y^b \phi(t,z)dz f(y)\partial_t L_t^y \, dy.$$

We can change the order of integration to get the final integral of (5.13). *Q.E.D.*

§6. A Formal Look at Some Higher Order Formulae

Suppose $\Phi_i(t,x)$ is an identifiable process of the form

$$(6.1) \qquad \Phi_i(t,x) = \Phi_i(t,a) + \int_a^x \phi_i(t,y)\partial_y L_t^y + \int_a^x \psi_i(t,y)dy$$

for $t \geq 0$, $x \geq a$, and $1 \leq i \leq n$, where ϕ_i and ψ_i are identifiable processes. It will be enough for our purposes to suppose ϕ_i and ψ_i are bounded, continuous, and in H. Let f_i be bounded and continuous and for each i, let $d\Lambda_i(t,x)$ be either $f_i(B_t)dH_t^x$ or dL_t^x (the choice may depend on i) and consider the multiple Stieltjes integral

$$(6.2) \quad I(x) = \int\cdots\int_{0 \leq s_1 \leq \cdots \leq s_n} \Phi_1(s_1,x)\cdots\Phi_n(s_n,x)\partial_{s_1}\Lambda_1(s_1,x)\cdots\partial_{s_n}\Lambda_n(s_n,x).$$

We claim that such an integral satisfies

$$(6.3) \qquad I(b) - I(a) = \iint_{R_b - R_a} p(t,x)\,dL_t^x + \int_a^b q(x)\,dx$$

where $p \in H$ and q is E_x-adapted. We have proved this for $n = 1$ in §5 (Theorem 5.2 and Corollary 5.3). We will treat the case $n > 1$ here. We will treat this purely formally. That is, we shall give a purely formal derivation of a few of the necessary formulae. Because of the complexity of the formulae, we will limit ourselves to the case $n = 2$, but it will be clear that the method extends equally well — or equally badly — to the general case.

We will proceed by discretization: we approximate the integral in (6.2) by Riemann sums, which we manipulate until we can recognize the corresponding sums approximating the integrals in (6.3). (The formulae in §5 can also be derived by this method. This is of course how they were originally found.)

The following discussion can be skipped at first reading. The only parts we will refer to later are equations (6.4) - (6.6). There are two tricky points in the discretization which we shall eventually bypass, but which we should discuss first. We can approximate $\iint \phi\,dL$ by the integrals of simple functions as follows: Let $x_0 < x_1 < x_2 < \cdots$, where $x_{i+1} - x_i \le \varepsilon$ for all i, and for each i, let $0 = S_{i0} \le S_{i1} \le S_{i2} \le \cdots$ be E_{x_i}-identifiable times such that $H^{x_i}_{S_{ij+1}} - H^{x_i}_{S_{ij}} \le \varepsilon$. We can do this by taking $S_{ij+1} = \inf\{t : H_t^{x_i} \ge H^{x_i}_{S_{ij}} + \varepsilon\}$, for example. This gives us a partition of the plane into special rectangles. Since any E_{x_i}-identifiable time is $E_{x_{i+1}}$-identifiable (see the remarks following Proposition 2.1) we can assume that for each i, the partition $\{S_{ij}\}$ is contained in $\{S_{i+1\,j}\}$. However, we cannot assume that the two partitions are equal, since $S_{i+1\,j}$ may fail to be $\mathcal{E}_{x_i}$-identifiable. This gives us a partition of the half-plane into special rectangles which looks something like this.

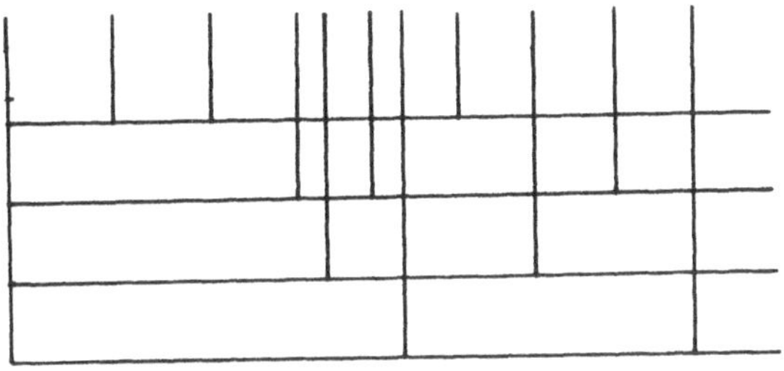

We can then approximate ϕ by a simple function $\hat{\phi}$ which is constant on each of these special rectangles, let the partition get finer and finer, and get $\iint \phi\, dL$ as the limit of $\iint \hat{\phi}\, dL$, that is, as a limit of Riemann sums. However, having chosen our partitions, we will also want to use them to calculate the two one-parameter integrals $\int \phi(t,x)\,\partial_x L_t^x$ and $\int \phi(t,x_i)\,\partial_t L_t^{x_i}$. There is no trouble with the former, since $\phi(t,x)$ is adapted to $\mathcal{E}_x \vee \sigma\{L_t^y,\ y \le x\}$ so that the Riemann sums $\sum_i \phi(t,x_i)(L_t^{x_{i+1}} - L_t^{x_i})$ are the classical approximating sums to the stochastic integral. The latter Stieltjes integral is going to be approximated by $\sum_j \phi(S_{ij},x_i)(L_{S_{ij+1}}^{x_i} - L_{S_{ij}}^{x_i})$. But now the mesh of the partition S_{ij} does not go to zero. (Indeed $B_{S_{ij}} \le x_i$ since S_{ij} is $\mathcal{E}_{x_i}$-identifiable, so that if there is an excursion *above* x_i from time s to time t, there can't be any S_{ij} in (s,t), so the mesh is at least $t-s$.) However, we are saved by two facts. First, $L_{S_{ij+1}}^{x_i} - L_{S_{ij}}^{x_i} \to 0$ uniformly on compacts as $\varepsilon \to 0$, and second, $\phi(\cdot,x_i)$, being identifiable, is constant on each excursion above x_i (Corollary A4 of the appendix). So in fact the Riemann sum will converge to the integral.

Let Δ_i indicate the ith increment in x and δ_{ij} indicate the jth increment in t at level x_i. For instance

$$\Delta_i L_t^x = L_t^{x_{i+1}} - L_t^{x_i} \quad \text{and} \quad \delta_{ij} L_t^x = L_{S_{ij+1}}^{x_i} - L_{S_{ij}}^{x_i}.$$

Then $\Delta_i \delta_{ij} L = \delta_{ij} \Delta_i L$ is the "rectangular" increment

$$L_{S_{ij+1}}^{x_{i+1}} - L_{S_{ij}}^{x_{i+1}} - L_{S_{ij+1}}^{x_i} + L_{S_{ij}}^{x_i}.$$

Consulting Theorem 2.5 and equations (2.8), we see that these increments satisfy

(6.4) $$E\{(\Delta_i \delta_{ij} L)(\Delta_k \delta_{k\ell} L)\} = 0 \quad \text{unless } i = k \text{ and } j = \ell\,;$$

(6.5) $$E\{(\Delta_i \delta_{ij} L)^2\} = 4\, E\{\int_{x_i}^{x_{i+1}} (L_{S_{ij+1}}^y - L_{S_{ij}}^y)\, dy\}.$$

This last quantity is approximately $4\Delta_i x\, E\{\delta_{ij} L^{x_i}\}$. Furthermore, as $\varepsilon \to 0$, the partition gets finer and finer, and one can show that

(6.6) $$\sum_{i,j: S_{ij} \le N} [(\Delta_i \delta_{ij} L)^2 - 4(\delta_{ij} L^{x_i}) \Delta_i x] \to 0.$$

The principal difficulty in actually using the above approximation procedure is not that the partitions are random, but rather that the partitions of the time-axis depend on the level x_i. This complicates the notation enough so that a careful use of it would make the following derivations incomprehensible. So we shall simply ignore all the above considerations and use the usual non-random partitions in what follows.

Fix $a < b$ and N, and let $a = x_0 < x_1 < \cdots < x_N = b$, and let $0 = t_0 < t_1 < t_2 < \cdots$ be partition of $[a,b]$ and $[0,\infty)$ respectively. The partition along the t-axis does not depend on the level x_i. We will use Δ_i for increments in x and δ_j for increments in t. We will explicitly indicate the variables when there might be some ambiguity, and suppress them if not. Thus, for instance,

$$\Delta_i L_t = L_t^{x_{j+1}} - L_t^{x_j}, \quad \delta_j L^x = L^x_{t_{j+1}} - L^x_{t_j},$$

and

$$\Delta_i \delta_j L = \delta_j \Delta_i L = L_{t_{j+1}}^{x_{i+1}} - L_{t_{j+1}}^{x_i} - L_{t_j}^{x_{i+1}} + L_{t_j}^{x_i}.$$

In order to handle the various increments of L and their products when they occur in Riemann sums, we construct the following multiplication table. Refer to (6.4) - (6.6) to see where it comes from. Let $s < t$.

	Δx	ΔL_s	ΔL_t	$\delta\Delta L_s$	$\delta\Delta L_t$
Δx	0	0	0	0	0
ΔL_s		$4L_s\Delta x$	$4L_s\Delta x$	0	0
ΔL_t			$4L_t\Delta x$	$4\delta L_s\Delta x$	0
$\delta\Delta L_s$				$4\delta L_s\Delta x$	0
$\delta\Delta L_t$					$4\delta L_t\Delta x$

Let Φ_1 and Φ_2 be given by (6.1). Define I(x) by

$$I(x) = \int_0^\infty \int_s^\infty \Phi_1(s,x)\Phi_2(t,x)\partial_t L_t^x \, \partial_s L_s^x. \tag{6.7}$$

We will show that I satisfies (6.3) with

$$p(u,x) = \int_u^\infty \int_s^\infty \Phi_1(s,x)\Phi_2(t,x)\partial_t L_t^x \partial_s L_s^x + \int_u^\infty \int_0^t \Phi_1(s,x)\Phi_2(t,x)\partial_s L_s^x \partial_t L_t^x$$
$$+ \Phi_2(u,x) \int_0^u \Phi_1(s,x)\partial_s L_s + \Phi_1(u,x) \int_u^\infty \Phi_2(t,x)\partial_t L_t^x \tag{6.8}$$

and

$$(6.9)\qquad q(x) = \int_0^\infty\int_s^\infty [4\phi_1(s,x)\phi_2(t,x)L_s^x + 4\Phi_1(s,x)\Phi_2(t,x)$$

$$+ \psi_1(s,x)\Phi_2(t,x) + \Phi_1(s,x)\psi_2(t,x)]\partial_t L_t^x \partial_s L_s^x .$$

REMARKS. It is not difficult to verify that p is identifiable and q is adapted. However, p is not in H as it is defined, since p does not necessarily vanish on S^-. (See §2 for the definitions of H, S^-, S^0 and S^+.) However, we can replace p by $\hat{p} = pI_{S^+}$ and $q(x)$ by $\hat{q}(x) = q(x) + 2p(0,x)I_{\{x<0\}}$. Then $\hat{p}$ vanishes on S^- and by (2.10)

$$\iint p\, dL + \int q\, dx = \iint \hat{p}\, dL + \int \hat{q}\, dx.$$

Let us write $I(b) - I(a) = \sum_{i=0}^{N-1} \Delta_i I$, and consider a typical $\Delta_i I$. We will fix i, so let $x = x_i$ and $y = x_{i+1}$, and omit the subscript i. Then

$$\Delta I(x) = \int_0^\infty\int_s^\infty \Delta\Phi_1(s,x)\Phi_2(t,y)\partial_t L_t^y \partial_s L_s^y + \int_0^\infty\int_s^\infty \Phi_1(s,x)\Delta\Phi_2(t,x)\partial_t L_t^y \partial_s L_s^y$$

$$+ \int_0^\infty\int_s^\infty \Phi_1(s,x)\Phi_2(t,x)(\partial_t \Delta L_t^x)\partial_s L_s^y + \int_0^\infty\int_s^\infty \Phi_1(s,x)\Phi_2(t,x)\partial_t L_t^x(\partial_s \Delta L_s^x)$$

$$\overset{\text{def}}{=} \Delta J_1(x) + \Delta J_2(x) + \Delta J_3(x) + \Delta J_4(x).$$

We will look at each of the ΔJ. Now $\partial_u L_u^y = \partial_u L_u^x + \partial_u \Delta L_u^x$ and, by (6.1), we have, approximately, that for any u, $i = 1, 2$,

$$\Delta\Phi_i(u,x) \sim \phi_i(u,x)L_s^x + \psi_i(u,x)\Delta x.$$

Thus

$$(6.10)\quad \Delta J_1 = \int_0^\infty\int_s^\infty (\phi_1(s,x)\Delta L_s^x + \psi_1(s,x)\Delta x)(\Phi_2(t,x) + \phi_2(t,x)\Delta L_t^x$$

$$+ \psi_2(t,x)\Delta x)\partial_s(L_s^x + \Delta L_s^x)\partial_t(L_t + \Delta L_t^x).$$

If we multiply this integral out and consult our multiplication table, we see that all but three of the integrands can be neglected. Since the only space variable appearing in (6.5) is x, we suppress it below.

$$(6.11) \quad \Delta J_1 = \int_0^\infty \int_s^\infty \phi_1(s)\Delta L_s \Phi_2(t)\partial_s L_s \partial_t L_t + \int_0^\infty \int_s^\infty \phi_1(s)\phi_2(t)(\Delta L_s \Delta L_t)\partial_s L_s \partial_t L_t$$

$$+ \int_0^\infty \int_s^\infty \psi_1(s)\Phi_2(t)\partial L_s \partial L_t \Delta x.$$

Discretize the first integral, writing $\Delta L_{t_m} = \sum_{i=0}^{m-1} \partial_i \Delta L.$

$$\sum_{m=0}^\infty [\sum_{n=m}^\infty \sum_{i=0}^{m-1} \phi_1(t_m)\Phi_2(t_n)\partial_i \Delta L \partial_m L]\partial_n L$$

$$= \sum_{i=0}^\infty [\sum_{m=i+1}^\infty \sum_{n=m}^\infty \phi_1(t_m)\Phi_2(t_n)\partial_m L \partial_n L]\partial_i \Delta L.$$

The sum in brackets is the Riemann sum for a double integral, so this is approximately

$$\sum_j (\int_{t_j}^\infty \int_s^\infty \Phi_1(s)\ \Phi_2(t)\ \partial_s L_s \partial_t L_t)\delta_j \Delta L.$$

Since $\Delta L_s \Delta L_t \sim 4L_s \Delta x$ according to our table, the second integral becomes

$$4 \int_0^\infty \int_s^\infty \phi_1(s) L_s \phi_2(t)\partial_s L_s \partial_t L_t\ \Delta x.$$

Putting this together (remember we are suppressing the x)

$$(6.12) \qquad \Delta J_1 = \sum_i \int_{t_j}^\infty \int_s^\infty \phi_1(s)\Phi_2(t)\partial_t L_t\ \partial_s L_s\ \delta_j \Delta L_{t_j}$$

$$+ \Big[\int_0^\infty \int_s^\infty (4\phi_1(s)L_s\phi_2(t) + \psi_1(s)\Phi_2(t))\partial_t L_t \partial_s L_s\Big]\Delta x.$$

We do the same for ΔJ_2. We continue to suppress the x.

$$(6.13) \qquad \Delta J_2 = \int_0^\infty \int_s^\infty \Phi_1(s)(\phi_2(t)\Delta L_t + \psi_2(t)\Delta x)\partial_t(L_t + \Delta L_t)\partial_s(L_s + \Delta L_s).$$

Multiply out and keep the non-negligible terms:

$$\sim \int_0^\infty \int_s^\infty \Phi_1(s)\phi_2(t)\Delta L_t \partial_t L_t \partial_s L_s + \int_0^\infty \int_s^\infty \Phi_1(s)\phi_2(t)(\Delta L_t \partial_s \Delta L_s)\partial_t L_t$$

$$+ \int_0^\infty \int_s^\infty \Phi_1(s)\psi_2(t)\partial_t L_t \partial_s L_s \Delta x.$$

Discretize the first integral and change order as before:

$$\sum_{n=0}^\infty \sum_{n=m}^\infty \sum_{i=0}^{n-1} \Phi_1(t_m)\phi_2(t_n)\partial_i \Delta L \; \partial_n L \partial_m L$$

$$= \sum_{i=0}^\infty \Big[\sum_{n=i+1}^\infty \sum_{m=0}^{n} \Phi_1(t_m)\phi_2(t_n)\partial_n L \partial_m L\Big]\partial_i \Delta L$$

$$\sim \sum_{i=0}^\infty \Big[\int_{t_{i+1}}^\infty \int_0^t \Phi_1(s)\phi_2(t)\partial_s L_s \partial_t L_t\Big]\partial_i \Delta L.$$

In the second integral, $\Delta L_t \partial_s L_s \sim 4\partial_s L_s \Delta x$, so it becomes

$$4\int_0^\infty \int_s^\infty \Phi_1(s)\phi_2(t)\partial_s L_s \partial_t L_t \Delta x.$$

Thus

$$(6.14) \qquad \Delta J_2 \approx \sum_{i=0}^\infty \int_{t_{i+1}}^\infty \int_0^t \Phi_1(s)\phi_2(t)\partial_s L_s \partial_t L_t \; \partial_i \Delta L$$

$$+ \int_0^\infty \int_0^S \big(4\Phi_1(s)\phi_2(t) + \Phi_1(s)\psi_2(t)\big)\partial_t L_t \partial_s L_s \; \Delta x.$$

Move on to ΔJ_3.

$$\Delta J_3 = \int_0^\infty \int_s^\infty \Phi_1(s)\Phi_2(t)\partial_t \Delta L_t \partial_s (L_s + \Delta L_s)$$

$$= \int_0^\infty \int_s^\infty \Phi_1(s)\Phi_2(t)\partial_t \Delta L_t \partial_s L_s + \int_0^\infty \int_s^\infty \Phi_1(s)\Phi_2(t)(\partial_t \Delta L_t)(\partial_s \Delta L_s).$$

Interchange the order in the first integral. In our discrete version, it is

$$\sim \sum_j \left(\int_0^{t_j} \Phi_1(s)\partial_s L_s\right)\Phi_2(t)\partial_j \Delta L.$$

Discretize the second integral:

$$\sum_{m=0}^\infty \sum_{n=m}^\infty \Phi_1(t_m)\Phi_2(t_n)\delta_n \Delta L \partial_m \Delta L.$$

There is a non-negligible contribution from the terms $m = n$, for $(\partial_m \Delta L)^2 \sim 4\partial_m L\Delta x$, leading to $4\sum_{m=0}^\infty \Phi_1(t_m)\Phi_2(t_m)\partial_m L\Delta x$. Thus

$$(6.15) \qquad \Delta J_3 \approx \sum_j \int_0^{t_j} \Phi_1(s)\partial_s L_s \Phi_2(t_j)\delta_j \Delta L + 4\int_0^\infty \Phi_1(s)\Phi_2(s)\partial_s L_s \Delta x.$$

Finally, we have

$$(6.16) \qquad \Delta J_4 \approx \sum_j \Phi_1(t_j) \int_{t_j}^\infty \Phi_2(t)\partial_t L_t \; \delta_j \Delta L_t.$$

Now define p and q by (6.8) and (6.9) respectively. Note that p is identifiable and q is E_x-adapted. From (6.12) - (6.16) we see that

$$\Delta I = \sum_j p(t_j,x)\delta_j \Delta L + q(x)\Delta x,$$

so that — if we cease suppressing the index i —

$$I(b) - I(a) \approx \sum_{i,j} p(t_k,x_i)\delta_j\Delta_i L + \sum_j q(x_i)\Delta_i x$$

which we recognize as Riemann sums for the integrals

$$\iint_{R_b - R_a} p(t,x)dL_t^x + \int_a^b q(x)dx$$

as claimed.

We will derive one more such formula. Let f be a smooth function and let

$$I(x) = \int_0^\infty \int_s^\infty \Phi_1(s,x)\Phi_2(t,x)\partial_s L_s^x \, f(B_t)\partial_t H_t^x$$

$$= \int_{-\infty}^x \int_0^\infty \int_s^\infty \Phi_1(s,x)\Phi_2(t,x)\partial_t L_t^z \partial_s L_s^x \, f(z)dz.$$

We will show that I satisfies (6.3) with

(6.17) $$p(u,x) = \int_u^\infty \int_s^\infty \Phi_1(s,x)\Phi_2(t,x)f(B_t)\partial_t H_t^x \partial_s L_s^x$$

$$+ \int_u^\infty \int_0^t \Phi_1(s,x)\phi_2(t,x)f(B_t)\partial_t H_t^x \partial_s L_s^x + \Phi_1(u,x)\int_u^\infty \Phi_2(t,x)f(B_t)\partial_t H_t^x,$$

(6.18) $$q(x) = f(x)\int_0^\infty \int_s^\infty \Phi_1(s,x)\Phi_2(t,x)\partial_t L_t^x \partial_s L_s^x$$

$$+ \int_0^\infty \int_s^\infty (4\phi_1(s,x)\phi_2(t,x)L_s^x + \psi_1(s,x)\Phi_2(t,x))f(B_t)\partial_t H_t^x \partial_s L_s^x$$

$$+ \int \int^\infty \Phi_1(s,x)(4\phi_2(t,x) + \psi_2(t,x))f(B_t)\partial_t H_t^x \partial_s L_s^x.$$

Let $x = x_i$ and $y = x_{i+1}$ as before. Then

$$\Delta I(x) = \int_x^y \int_0^\infty \int_s^\infty \Phi_1(s,y)\Phi_2(t,y)\partial_t L_t^z \partial_s L_s^y f(z)dz$$

$$+ \int_{\infty}^{x}\int_{0}^{\infty}\int_{s}^{\infty} \Delta\Phi_1(s,x)\Phi_2(t,y)\partial_t L_t^z \partial_s L_s^y f(z)dz$$

$$+ \int_{-\infty}^{x}\int_{0}^{\infty}\int_{s}^{\infty} \Phi_1(s,x)\Delta\Phi_2(t,x)\partial_t L_t^z \partial_s L_s^y f(z)dz$$

$$+ \int_{-\infty}^{x}\int_{0}^{\infty}\int_{s}^{\infty} \Phi_1(s,x)\Phi_2(t,x)\partial_t L_t^z \partial_s \Delta L_s^x f(z)dz$$

$$\stackrel{\text{def}}{=} \Delta J_1 + \Delta J_2 + \Delta J_3 + \Delta J_4.$$

Now

$$\Delta J_1 \approx \int_{0}^{\infty}\int_{s}^{\infty} (\Phi_1(s,x) + \phi_1(s,x)\Delta L_s^x + \psi_1(s,x)\Delta x)(\Phi_2(t,x) + \phi_2(t,x)\Delta L_t^x$$

$$+ \psi_2(t,x)\Delta x)\partial_t L_t^x \partial_s (L_s^x + \Delta L_s) f(x)\Delta x.$$

Consulting the multiplication table, we see that there is only a single non-negligible term in all this:

$$\text{(6.19)} \qquad \Delta J_1 \approx \int_{0}^{\infty}\int_{s}^{\infty} \Phi_1(s,x)\Phi_2(t,x)\partial_t L_t^x \partial_s L_s^x f(x)\Delta x.$$

Let us again suppress the x whenever possible. Then

$$\Delta J_2 \approx \int_{-\infty}^{x}\int_{0}^{\infty}\int_{s}^{\infty} (\phi_1(s)\Delta L_s + \psi_1(s)\Delta x)(\Phi_2(t) + \phi_2(t)\Delta L_t$$

$$+ \psi_2(t)\Delta x)\partial_t L_t^z \partial_s (L_s^x + \Delta L_s) f(z)dz$$

$$\approx \int_{-\infty}^{x}\int_{0}^{\infty}\int_{s}^{\infty} \phi_1(s)\Delta L_s \Phi_2(t)\partial_t L_t^z \partial_s L_s^x f(z)dz$$

$$+ \int_{-\infty}^{x}\int_{0}^{\infty}\int_{s}^{\infty} \phi_1(s)\phi_2(t)\Delta L_s \Delta L_t \partial_t L_t^z \partial_s L_s^x f(z)dz$$

$$+ \int_{-\infty}^{x}\int_{0}^{\infty}\int_{s}^{\infty} \psi_1(s)\Phi_2(t)\partial_t L_t^z \partial_s L_s^x f(z)dz\Delta x$$

+ negligible terms.

Discretize the first integral — or at least the part over s and t for fixed z:

$$\sum_{m=0}^{\infty}\sum_{n=m}^{\infty}\sum_{j=0}^{m-1} \phi_1(t_m)\Phi_2(t_n)\partial_m L(\delta_j\Delta L)f(z)\delta_n L_t^z$$

$$= \sum_{j=0}^{\infty}\Big[\sum_{m=i+1}^{\infty}\sum_{n=m}^{\infty} \phi_1(t_m)\Phi_2(t_n)\partial_m L^x f(z)\delta_n L^z\Big](\delta_j\Delta L).$$

We recognize the term in brackets as a Riemann sum. We conclude that the first integral is approximately

$$\sum_{j=0}^{\infty}\Big[\int_{-\infty}^{x}\int_{t_{i+1}}^{\infty}\int_{s}^{\infty} \phi_1(s)\Phi_2(t)\partial_t L_t^z f(z)\partial_s L_s dz\Big]\delta_j\Delta L.$$

In the second integral, $\Delta L_s \Delta L_t \sim 4L_s\Delta x$, so that this integral corresponds to:

$$4\int_{-\infty}^{x}\int_{0}^{\infty}\int_{s}^{\infty} \phi_1(s)\phi_2(t)L_s^x\partial_t L_t^z\partial_s L_s^x f(z)dz\Delta x.$$

Thus, in terms of H_t^x,

$$\text{(6.20)}\qquad \Delta J_2 \approx \sum_{i=0}^{\infty}\int_{t_{i+1}}^{\infty}\int_{s}^{\infty}\phi_1(s)\Phi_2(t)f(B_t)\partial_t H_t^x\partial_s L_s^x\partial_i\Delta L$$

$$+\int_{0}^{\infty}\int_{s}^{\infty}(4\phi_1(s)\phi_2(t)L_s^x + \psi_1(s)\Phi_2(t))f(B_t)\partial_t H_t^x\partial_s L_s^x$$

$$\Delta J_3(t) \approx \int_{-\infty}^{x}\int_{0}^{\infty}\int_{s}^{\infty}\Phi_1(s)(\phi_2(t)\Delta L_t + \psi_2(t)\Delta x)\partial_t L_t^z f(z)\partial_s(L_s + \Delta L_s)dz.$$

Expanding and keeping the non-negligible terms:

$$\approx \int_{-\infty}^{x}\int_{0}^{\infty}\int_{s}^{\infty}\Phi_1(s)\phi_2(t)\Delta L_t\partial_t L_t^z\partial_s L_s f(z)dz$$

$$+ \int_{-\infty}^{x} \int_0^\infty \int_s^\infty \Phi_1(s)\phi_2(t)\Delta L_t \partial_t L_t^z \partial_s \Delta L_s f(z)dz$$

$$+ \int_{-\infty}^{x} \int_0^\infty \int_s^\infty \Phi_1(s)\psi_2(t)\partial_t L_t^z \partial_s L_s f(z)dz\Delta x.$$

Fix z and discretize the first integral. It is approximately

$$\sum_{m=0}^{\infty} \sum_{n=m}^{\infty} \sum_{j=0}^{n-1} \Phi_1(t_m)\phi_2(t_n)\delta_j \Delta L^x \delta_n L^z \delta_m L^x f(z)$$

$$= \sum_{j=0}^{\infty} \Big[\sum_{n=j+1}^{\infty} \sum_{m=0}^{n} \Phi_1(t_m)\phi_2(t_n)\delta_m L^x \delta_n L^z \Big] f(z)\delta_j \Delta L.$$

We recognize the term in square brackets as an approximation to a double integral. Integrate this over z to see that the first integral is approximately

$$\sum_{j=0}^{\infty} \Big[\int_{-\infty}^{x} \int_{t_{y+1}}^{\infty} \int_0^{t_j} \Phi_1(s)\phi_2(t)\partial_t L_t^z \partial_s L_s^x f(z)dz \Big] \delta_j \Delta L.$$

In the second integral, as $t > s$, $\Delta L_t \partial_s L_s \sim 4\partial_s L_s \Delta x$ so that we get

$$4 \int_{-\infty}^{x} \int_0^\infty \int_s^\infty \Phi_1(s)\phi_2(t)\partial_t L_t^z \partial_s L_s^x f(z)dz\Delta x = 4\int_0^\infty \int_s^\infty \Phi_1(s)\phi_2(t) f(B_t)\partial_t H_t^x \partial_s L_s^x \Delta x.$$

Thus

$$\text{(6.21)} \qquad J_3 \approx \sum_{j=0}^{\infty} \Big[\int_{t_{j+1}}^{\infty} \int_0^{t} \Phi_1(s)\phi_2(t) f(B_t)\partial_t H_t^x \partial_s L_s^x \Big] \delta_j \Delta L$$

$$+ \Big[\int_0^\infty \int_s^\infty \Phi_1(s)(4\phi_2(t) + \psi_2(t) f(B_t)\partial_t H_t^x \partial_s L_s^x \Big] \Delta x.$$

Finally, ΔJ_4 is approximately

$$\int_{-\infty}^{x} \int_0^\infty \int_s^\infty \Phi_1(s,x)\Phi_2(t)\partial_t L_t^z \partial_s \Delta L_s^x f(z)dz \approx \sum_j \int_{-\infty}^{x} \int_{t_i}^{\infty} \Phi_1(t_j)\Phi_2(t)\partial_t L_t^z f(z)dz\delta_j \Delta L.$$

Thus

$$\Delta J_4 \approx \sum_j \Big[\Phi(t_j)\int_{t_i}^{\infty} \Phi_2(t) f(B_t)\partial_t H_t^x\Big]\delta_j \Delta L. \tag{6.22}$$

Now define p and q by (6.17) and (6.18). Note that p is identifiable and q is E_x-adapted. From (6.19) - (6.22) we see that

$$\Delta I \approx \sum_j p(t_j, x)\delta_j \Delta L + q(x)\Delta x$$

so that

$$I(b) - I(a) \approx \sum_{i,j} p(t_j, x_i)\delta_j \Delta_i L + \sum_i q(x_i)\Delta_i x$$

$$\approx \iint_{R_b - R_a} p(t,x)\, dL_t^x + \int_a^b q(x)\, dx$$

as claimed.

§7. Martingale Representations

Let $M^2(E)$ be the class of all martingales $\{M_x, E_x,\ x \in \mathbb{R}\}$ such that

$$\|M\|_m^2 \overset{\text{def}}{=} \sup_x E\{M_x^2\} < \infty.$$

The purpose of this section is to prove the following.

THEOREM 7.1. Let $M \in M^2(E)$. Then there exists a unique $\phi \in H$ and a constant $M_{-\infty}$ such that for all real x

$$M_x = M_{-\infty} + \phi \cdot L(x). \tag{7.1}$$

LEMMA 7.2. Let $M \in M^2(E)$. A necessary and sufficient condition that (7.1) hold is that there exists $\phi \in H$ such that for each real $a < b$

(7.2) $$M_b - M_a = \iint_{R_b - R_a} \phi dL.$$

PROOF. The necessity is clear. Conversely, if (7.2) holds, $M_{-\infty} = \lim_{x \to -\infty} M_x$ is $\bigcap_x E_x$-measurable and therefore, by Proposition 2.4, is constant. Let $a \to -\infty$ in (7.2). The right-hand side tends to $\iint_{R_b} \phi dL = \phi \cdot L(b)$.

Let $R^2(E)$ be the set of $M \in M^2(E)$ which satisfies (7.1), and let $R_0^2(E)$ be the set of $M \in R^2(E)$ for which $M_{-\infty} = 0$.

LEMMA 7.3. $R^2(E)$ is closed in $M^2(E)$.

PROOF. Since $M_{-\infty}$ is constant for any $M \in M^2(E)$, it is enough to consider the case where $M_{-\infty} = 0$. Now if $M = \phi \cdot L$, $\|M\|_m = \|\phi\|_H$. It follows that the map $\phi \to \phi \cdot L$ of H onto $R_0^2(E)$ is an isometry. Since H is closed, so is $R_0^2(E)$. *Q.E.D.*

In order to prove Theorem 7.1, it is sufficient to show that (7.2) holds for a dense subset of $M^2(E)$.

Thus, it is enough to verify (7.2) for martingales of the form

$$M_x = E\{f_1(B_{t_1}) \cdots f_n(B_{t_n}) | E_x\}$$

for bounded smooth f_i and $t_1 < \cdots < t_n$, or even for martingales of the form

(7.3) $$M_x = E\{\int_0^\infty \cdots \int_0^\infty \prod_{j=1}^n e^{-\lambda_j t_j} f_j(B_{t_j}) dt_1 \cdots dt_n | E_x\}.$$

If $S_1, \ldots, S_n$ are independent exponential random variables which are independent of B and which have parameters $\lambda_1, \ldots, \lambda_n$ respectively, then (7.3) is equal to

(7.4) $$E\{f_1(B_{S_1})\cdots f_n(B_{S_n})|\mathcal{E}_x\}$$

Williams [4] has pointed out that it is sufficient to consider a slightly different class of martingales, namely

(7.5) $$N_x = E\{f_1(B_{S_1})f_2(B_{S_1+S_2})\cdots f_n(B_{S_1+\cdots+S_n})|\mathcal{E}_x\}.$$

These are more suitable for induction arguments than those of the form (7.4), and Williams has shown how to get explicit expressions for these. (The article [4] treats Bessel processes rather than Brownian motions, but the calculations are easily modified to handle our situation. In fact, our situation is slightly easier, since the Bessel process has a finite minimum, whereas Brownian motion does not.)

The idea of our proof is to use Williams' explicit expressions, showing that each of these is representable by means of our Green's formulas.

We consider the case $n = 1$ first. The ideas involved in the cases $n = 1$ and $n > 1$ are the same, so we will treat the former case carefully, and then just indicate how the general case goes.

Let S be exponential (λ), independent of $\{B_t, t \geq 0\}$, let f be a smooth function of compact support, and put

$$M_x = E\{f(B_S)|\mathcal{E}_x\}.$$

Let $$A_1(x) = \{\omega: S < T_{(-\infty,x]}\}$$

$$A_2(x) = \{\omega: S \geq T_{(-\infty,x]}, B_S > x\}$$

$$A_3(x) = \{\omega: S \geq T_{(-\infty,x]}, B_S \leq x\}.$$

$$\text{Let } p_x(t,y) = \begin{cases} (2\pi t)^{-\frac{1}{2}}\left(e^{\frac{-y^2}{2t}} - e^{\frac{(y+x)^2}{2t}}\right) & \text{if } x \le y \\ 0 & \text{if } x > y \end{cases}$$

$$\gamma = \left(\frac{\lambda}{8}\right)^{\frac{1}{2}}, \quad C_\lambda f(x) = (2\lambda)^{3/2} \int_x^\infty e^{\lambda(x-y)} f(y)dy$$

$$\text{and} \quad g(x,\lambda) = P\{S > T_{(-\infty,x]}\} = \begin{cases} 2\int_0^\infty \lambda e^{-\lambda t} P\{B_t < x\}dt & \text{if } x < 0 \\ 1 & \text{if } x \ge 0 \end{cases}.$$

Then Williams showed that

$$(7.6) \quad E\{f(B_S),\ A_1(x)|E_x\} = \lambda \int_0^\infty \int_{-\infty}^\infty p_x(t,y)f(y)e^{-\lambda t}dtdy \quad \stackrel{\text{def}}{=} I_1(x);$$

$$(7.7) \quad E\{f(B_S),\ A_2(x)|E_x\} = g(x,\lambda)C_\lambda f(x)\int_0^\infty e^{-\lambda H_t^x - \gamma L_t^x}\partial_t L_t^x \stackrel{\text{def}}{=} I_2(x);$$

$$(7.8) \quad E\{f(B_S),\ A_3(x)|E_x\} = g(x,\lambda)\int_0^\infty e^{-\lambda H_t^x - \gamma L_t^x} f(B_t)\partial_t H_t^x \stackrel{\text{def}}{=} I_3(x).$$

Neither I_1, I_2 nor I_3 is a martingale, since the sets $A_i(x)$ depend on x. However, their sum is the martingale $M_x = E\{f(B_S)|E_x\}$. We will show that there are $\phi_i \in H$ and processes V_i of finite variation, i = 1,2,3, such that

$$I_i(x) = \phi_i \cdot L(x) + V_i(x), \quad i = 1,2,3.$$

If $\phi = \phi_1 + \phi_2 + \phi_3$, then $M_x - \phi \cdot L = V_1 + V_2 + V_3$ is a continuous martingale of finite variation, and hence is a constant, which implies (7.1).

Put $e_\lambda(t,x) = e^{-\lambda H_t^x - \gamma L_t^x}$ and set

$$J_2(x) = \int_0^\infty e_\lambda(t,x)\partial_t L_t^x; \quad J_3(x) = \int_0^\infty e_\lambda(t,x)f(B_t)\partial_t H_t^x.$$

By Ito's formula, if a < x

$$e_\lambda(t,x) = e_\lambda(t,a) - \gamma\int_a^x e_\lambda(t,y)\partial_y L_t^y + (2\gamma^2 - \lambda)\int_a^x e_\lambda(t,y)L_t^y\,dy.$$

According to Theorem 5.2

$$(7.9) \qquad J_2(b) - J_2(a) = \int_a^b \int_0^\infty ((2\gamma^2 - \lambda)L_t^y - 2\gamma)e_\lambda(t,y)\partial_t L_t^y dy$$

$$+ \iint_{R_b - R_a} (e_\lambda(t,y) - \gamma\int_t^\infty e_\lambda(s,y)\partial_s L_s^y)dL_t^y$$

and by Corollary 5.3

$$(7.10) \qquad J_3(b) - J_3(a) = -\gamma \iint_{R_b - R_a} [\int_a^x \int_t^\infty e_\lambda(s,x)\delta(y)\partial_s L_s^y dy]\, dL_t^x$$

$$+ \int_a^b [\int_0^\infty e_\lambda(t,x)\delta(x)\partial_t L_t^x + (2\gamma^2 - \lambda)\int_a^x \int_0^\infty e_\lambda(t,x)L_t^x \partial_t L_t^y \delta(y)dy]\, dx.$$

(In fact, we can't apply Theorem 5.2 and Corollary 5.3 directly above, since $e_\lambda(t,x)$ does not have compact support in t. We apply them instead to $e_\lambda(t,x)I_{\{t \le T_{-N}\}}$, where $-N < a$, and let $N \to \infty$. Since e_λ is bounded and tends to zero exponentially as $t \to \infty$, we can pass to the limit to get (7.9) and (7.10).)

Rewrite (7.9) and (7.10) to simplify notation:

$$J_i(b) - J_i(a) = \iint_{R_b - R_a} \psi_i(t,x)dL_t^x + \int_a^b \eta_i(x)dx, \quad i = 2,3$$

and note that $I_i(x) = g_i(x)J_i(x)$, where $g_2(x) = g(x,\lambda)C_\lambda f(x)$ and $g_3(x) = g(x,\lambda)$. Now g_i is differentiable, so by (3.7)

$$I_i(b) - I_i(a) = \iint_{R_b - R_a} g_i(x)\psi_i(t,x)dL_t^x + \int_a^b (g_i(x)\eta_i(x) + J_i(x)g_i'(x))dx$$

$$\overset{\text{def}}{=} \iint_{R_b - R_a} \hat{\phi}_i dL + \int_a^b \hat{\nu}_i(x)dx,$$

which holds for i = 2 and 3, and for i = 1 as well with $\hat{\phi}_1 \equiv 0$.

Note that $\hat{\phi}_i(0,x)$ does not necessarily vanish, so $\hat{\phi}_i \in \hat{H}$, but not in H. Thus let

$$\phi_i(t,x) = \hat{\phi}_i(t,x) I_{S^+}(t,x)$$

and
$$\nu_i(x) = \hat{\nu}_i(x) + 2\hat{\phi}(0,x) I_{\{x<0\}},$$

where S^+ is defined in §2. Then $\phi_i \in H$ and

$$\iint_{R_b - R_a} \phi_i dL + \int_a^b \nu_i dx = \iint_{R_b - R_a} \hat{\phi}_i dL + \int_a^b \hat{\nu}_i dx = I_i(b) - I_i(a).$$

This takes care of the case n = 1. Turning to the case n > 1, we note that, following Williams' induction argument [4], the martingale N_x in (7.5) can be written as a sum of terms of the form

$$g(x)\int_0^\infty \cdots \int_0^\infty h_1(L^x_{s_1}, H^x_{s_1}) \cdots h_m(L^x_{s_1+\cdots+s_m}, H^x_{s_1+\cdots+s_m})$$
$$\partial_{s_1}\Lambda^x_1(s_1) \cdots \partial_{s_m}\Lambda^x_m(s_1 + \cdots + s_m)$$

where $m \le n$, g and the h_i are smooth functions, and, for each i, $\partial_t \Lambda^x_i(t)$ equals either $\partial_t L^x_t$ or $f_i(B_t)\partial_t H^x_t$.

Let I(x) = g(x)J(x), where J(x) is the above multiple integral. By (6.3), we can write

$$J(b) - J(a) = \iint_{R_b - R_a} p(t,x) dL^x_t + \int_a^b q(x)dx$$

where $p \in H$ and q is E_x-adapted. By (3.7),

$$I(b) - I(a) = \iint_{R_b - R_a} g(x)p(t,x)dL_t^x + \int_a^b (J(x)g'(x) + q(x)g(x))dx.$$

Since N is a sum of such terms, there exists $\phi \in H$ and an adapted ν such that

$$N_b - N_a = \iint_{R_b - R_a} \phi(t,x)dL_t^x + \int_a^b \nu(x)dx.$$

But the right-hand side is a martingale in b, so that the bounded-variation term vanishes, i.e. $\nu = 0$. Thus N satisfies (7.2) and we are done.

Appendix: Identifiability

We will collect some results on identifiable processes and identifiable times here. Let us recall some notation.

$$H_t^x = \int_0^t I_{\{B_s \leq x\}} ds$$

$$\rho(t,x) = \inf\{s: H_s^x \geq t\}$$

$$Y_t^x = B_{\rho(t,x)}.$$

The process Y_t^x is a time-change of B which eliminates all excursions above x, and E_x is generated by $\{Y^x\}$. The natural parameter space of Y is $\mathbb{R}_+ \times \mathbb{R}$. We can also think of the parameter space of B as $\mathbb{R}_+ \times \mathbb{R}$, even though B does not depend on the space variable. Now let us consider the transformation Γ of $\mathbb{R}_+ \times \mathbb{R}$ into itself (which we think of as a transformation of "B-space" into "Y-space") given by

$$\Gamma(t,x) = (H_t^x, x).$$

The map will look something like this:

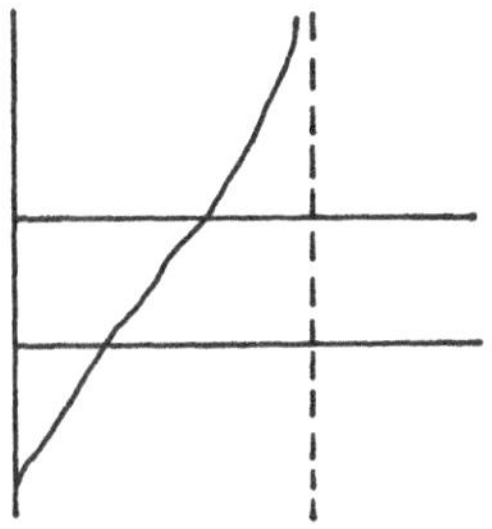

Let $\mathcal{P}$ be the σ-field of subsets of $\Omega \times \mathbb{R}_+ \times \mathbb{R}$ generated by all sets A of the form $A = (S,T] \times (x,y]$, where $x < y$ and $S \le T$ are $\mathcal{E}_x$-measurable random variables. We call $\mathcal{P}$ the class of *predictable sets* for Y.

THEOREM A1. $\mathcal{I}$ and $\mathcal{P}$ are isomorphic under the map Γ; that is, $\Gamma(\mathcal{I}) = \mathcal{P}$ and $\Gamma^{-1}(\mathcal{P}) = \mathcal{I}$, and for each $A \in \mathcal{I}$ and $C \in \mathcal{P}$, $\Gamma^{-1}\Gamma(A) \doteq A$ and $\Gamma\,\Gamma^{-1}(C) \doteq C$, where "$\doteq$" means "equality up to an evanescent set."

Before proving this, we need the following Lemma.

LEMMA A2. Let $\tau > 0$ be $\mathcal{E}_x$-identifiable. Then with probability one, for all $y \ge x$, $t \to H_t^y$ is strictly increasing at both τ^+ and τ^-. Moreover, there exist sequences σ_n and τ_n of $\mathcal{E}_x$-identifiable times such that with probability one $\sigma_n < \tau < \tau_n$, $\sigma_n \uparrow \tau$ and $\tau_n \downarrow \tau$.

PROOF. By Proposition 2.1, $B_\tau \le x$. On $\{B_\tau < x\}$, H_t^x is strictly increasing in a neighborhood of τ. On $\{B_\tau = x\}$, τ does not coincide with either the initial or final point of an excursion (Lemma 2.2). Thus τ is a limit from both left and right of t for which $B_t < x$, so for any small enough $\varepsilon > 0$, $H_{\tau-\varepsilon}^x < H_\tau^x < H_{\tau+\varepsilon}^x$. The same inequality holds with x replaced by any $y > x$, since for any $s < t$ and all ω, $H_t^y(\omega) - H_s^y(\omega) \ge H_t^x(\omega) - H_s^x(\omega)$.

Now let $\sigma_n = \inf\{t: H_t^x > H_\tau^x - \frac{1}{n}\} = \rho(H_\tau^x - \frac{1}{n}, x)$. This is E_x-identifiable by Corollary 2.2(e) and clearly satisfies $\sigma_n < \tau$, $\sigma_n \uparrow \tau$. The argument for $\tau_n = \rho(H_\tau^x + \frac{1}{n}, x)$ is similar.

PROOF (of Theorem A1). Let $T > 0$ be E_x-measurable. Then $\tau \overset{\text{def}}{=} \rho(T,x)$ is strictly positive and E_x-identifiable. By Lemma A2, τ is the unique point for which $H_\tau^x = T$. Moreover, we have simultaneously for all $y \geq x$ that $\tau = \rho(H_\tau^y, y)$.

Consequently, if $A = (\sigma,\tau] \times (x,y]$, where $0 < \sigma < \tau$ are E_x-identifiable and $x < y$, then $\Gamma(A) = \{(t,z): H_\sigma^z < t \leq H_\tau^z, x < z < y\}$, and $\Gamma^{-1}\Gamma(A) = A$. Similarly, if $A' = [0,\tau] \times (x,y]$, then $\Gamma^{-1}\Gamma(A') = A'$.

Let us verify that $\Gamma(A) \in P$. Partition $[x,y]$ into n equal segments by $x = x_0 < x_1 < \cdots < x_n = y$. Set $S_{nj} = H_\sigma^{x_j}$ and $T_{nj} = H_\tau^{x_j}$. Define

$$\Lambda_{\varepsilon n} = \{\omega: H_\tau^b - H_\tau^a < \varepsilon, \quad \text{all } x \leq a < b \leq y \ni b - a \leq (y-x)/n\}.$$

Let

$$D(n,j;\varepsilon) = (S_{nj} + \varepsilon, T_{nj} + \varepsilon] \times (x_j, x_{j+1}]$$

and

$$D(n,\varepsilon) = \bigcup_{j=0}^{n-1} D(n,j,\varepsilon).$$

Note that on $\Lambda_{\varepsilon n}$, since $\sigma \leq \tau$ we have $0 \leq H_\sigma^z - H_\sigma^{x_j} \leq H_\tau^z - H_\tau^{x_j} \leq \varepsilon$ for any $z \in [x_j, x_{j+1}]$, so that on $\Lambda_{\varepsilon n}$ we have for any $z \in (x,y]$

$$(H_\sigma^z + \varepsilon, H_\tau^z] \times \{z\} \subset D(n;\varepsilon) \subset (H_\sigma^z, H_\tau^z + \varepsilon] \times \{z\}.$$

Choose a sequence of $\varepsilon_k \to 0$ and $n_k \to \infty$ such that $\sum P\{\Lambda_{\varepsilon_k n_k}^c\} < \infty$; then

$$\Gamma(A) \doteq \limsup_{k \to \infty} D(n_k, \varepsilon_k),$$

which is certainly in P. A similar argument shows that $\Gamma((0,\tau] \times (x,y]) \in P$.

Now let R be the subset of I consisting of sets A with the property that $\Gamma(A)$ and $\Gamma(A^c)$ are in P, and both $A \doteq \Gamma^{-1}\Gamma(A)$ and $A^c \doteq \Gamma^{-1}\Gamma(A^c)$.

If $A = (\sigma,\tau] \times (x,y]$ is a special rectangle with $\sigma > 0$, then

$$A^c = \mathbb{R}_+ \times (-\infty,x] \cup \mathbb{R}_+ \times (y,\infty) \cup [0,\sigma] \times (x,y] \cup (\tau,\infty] \times (x,y].$$

We have just seen that $\Gamma(A) \in P$. Likewise, A^c is a union of special rectangles, so $\Gamma(A^c) \in P$. Moreover, $A = \Gamma^{-1}\Gamma(A)$ by our opening remarks and, by the same token, $A^c = \Gamma^{-1}\Gamma(A^c)$. Thus $A \in R$.

We claim R is a σ-field. It is closed under complementation by symmetry, and certainly contains ϕ. Then $\Gamma(\bigcup_n A_n) = \bigcup_n \Gamma(A_n) \in P$, and $\Gamma^{-1}\Gamma(\bigcup_n A_n) = \Gamma^{-1}(\bigcup_n \Gamma(A_n)) = \bigcup_n \Gamma^{-1}\Gamma(A_n) = \bigcup_n A_n$, since $A_n \in \mathbb{R}$. Turning to the complement, write $(\bigcup_n A_n)^c = \bigcap_n A_n^c$. Note that $\Gamma(\bigcap_n A_n^c) \subset \bigcap_n \Gamma(A_n^c)$. However, if $y \in \bigcap_n \Gamma(A_n^c)$, then $\Gamma^{-1}(y) \in \bigcap_n \Gamma^{-1}\Gamma(A_n^c) = \bigcap_n A_n^c$ since $A_n^c \in R$, so $y \in \Gamma(\bigcap_n A_n^c)$. This implies that $\Gamma(\bigcap_n A_n^c) = \bigcap_n \Gamma(A_n^c)$. Since $\Gamma(A_n^c) \in P$, all n, we have $\Gamma(\bigcap_n A_n^c) \in P$. Moreover, applying Γ^{-1}, we have

$$\Gamma^{-1}\Gamma(\bigcap_n A_n^c) = \bigcap_n \Gamma^{-1}\Gamma(A_n^c) = \bigcap_n A_n^c.$$

This shows that $\bigcup_n A_n \in R$, and verifies that R is a σ-field. Since R contains the generators of I, $R = I$.

To finish the proof, we need only show that $\Gamma^{-1}(P) \subset I$. Since Γ^{-1} preserves set operations, it is enough to show that $\Gamma^{-1}(C) \in I$ for sets C of the form

$$C = (S,T] \times (x,y]$$

where $0 < S < T$ are $\mathcal{E}_x$-measurable and $x < y$.

Partition $(x,y]$ into equal segments as before by $x = x_0 < x_1 < \cdots < x_n = y$. Let $\sigma_j = \rho(S,x_j)$, $\tau_j = \rho(T,x_j)$. Then $0 < \sigma_j \le \tau_j$ are $\mathcal{E}_{x_j}$-identifiable by Corollary 2.2(e). Let $A_{nj} = (\sigma_j,\tau_j] \times (x_j,x_{j+1}]$, $A_n = \bigcup_{j=0}^{n-1} A_{nj}$. Then $A_n \in \mathcal{I}$.

As before, let $\Lambda_{n\varepsilon} = \{\omega: H^b_{\tau_0} - H^a_{\tau_0} \le \varepsilon,\ \forall\ x \le a \le b \le y \ni b - a < (y-x)/n\}$. Now $\sigma_j \le \tau_j \le \tau_0$ for all j, so that on $\Lambda_{n\varepsilon}$, for any j, if $x_j \le z \le x_{j+1}$,

$$H^z_{\sigma_j} - H^{x_j}_{\sigma_j} \le H^z_{\tau_j} - H^{x_j}_{\tau_j} \le H^z_{\tau_0} - H^{x_j}_{\tau_0} \le \varepsilon .$$

It follows that on $\Lambda_{n\varepsilon}$

$$(S+\varepsilon,\ T] \times (x,y] \subset \Gamma(A_n) \subset (S,\ T+\varepsilon] \times (x,y].$$

Now A_n is a special rectangle, so $A_n = \Gamma^{-1}\Gamma(A_n)$, and

$$\Gamma^{-1}((S+\varepsilon,\ T] \times (x,y]) \subset A_n \subset \Gamma^{-1}((S,\ T+\varepsilon] \times (x,y]).$$

Now choose a sequence of $\varepsilon_k \to 0$ and $n_k \to \infty$ such that $\sum P\{\Lambda_{n_k\varepsilon_k}\} < \infty$, which we can do by the continuity of $z \to H^z_{\tau_0}$, to see that $\Gamma^{-1}(C) = \limsup_{k\to\infty} A_{n_k} \in \mathcal{I}$. This finishes the proof.

There are several immediate consequences of this result.

COROLLARY A3. If ϕ is identifiable, there exists a predictable ψ such that $\phi = \psi \circ \Gamma$ and $\psi = \phi \circ \Gamma^{-1}$ up to evanescent sets.

COROLLARY A4. Fix x. If α and β are the initial and final points of an excursion above x, and if ϕ is identifiable, then $t \to \phi(t,x)$ is constant on $[\alpha,\beta]$.

The fact that ϕ is constant on the closed interval follows from the fact that $t \to H_t^x$ is constant on any excursion above x, so that Γ^{-1} maps $[\alpha,\beta] \times \{x\}$ into a single point.

EXAMPLES. 1° $\phi(t,x) = I_{\{(t,x):B_t < x\}}$ is identifiable, but $I_{\{(t,x):B_t \le x\}}$ is not. This follows from the fact that $\phi = \psi \circ \Gamma$, where $\psi = I_{\{(t,x):Y_t^\gamma < x\}}$. However, in order for $I_{\{(t,x):B_t \le x\}}$ to be identifiable, $I_{\{(t,x):B_t = x\}}$ would have to be so. But this is not constant on excursions above x, and so violates Corollary A4.

2° Let ϕ be identifiable. Then, if the integrals exist, $\int_0^t \phi(s,x)\partial_s H_s^x$, $\int_t^\infty \phi(s,x)\partial_s H_s^x$, $\int_0^t \phi(s,x)\partial_s L_s^x$ and $\int_t^\infty \phi(s,x)\partial_s L_s^x$ all define identifiable functions.

Indeed, let $\psi(t,x) = \int_0^t \phi(\rho(s,x), x)ds$. Then $\int_0^t \phi(s,x)\partial_s H_s^x = \psi \circ \Gamma(t,x)$, while if $\eta(t,x) = \int_0^t \phi(\rho(s,x),x)\partial_s L^x_{\rho(s,x)}$ then $\int_0^t \phi(s,x)\partial_s H_s^x = \eta \circ \Gamma$. Both ψ and η are easily seen to be $\mathcal{P}$-measurable.

3° Let $\{X_x, x \in \mathbb{R}\}$ be predictable relative to $(\mathcal{E}_x)$ in the usual one-parameter sense. Then $\phi(x,t) = X_x$ is identifiable, for $\phi \circ \Gamma^{-1} \overset{\text{def}}{=} \phi(x,t)$ is certainly $\mathcal{P}$-measurable.

A criterion which is sometimes useful, and which in fact applies to the examples in 2° if ϕ is continuous, is the following.

PROPOSITION A5. Let $X = \{X(t,x): t \ge 0, x \in \mathbb{R}\}$ satisfy

(i) $(t,x) \to X(t,x)$ is a.s. continuous;

(ii) $X(\rho(t,x), x)$ is $\mathcal{E}_x$-measurable;

(iii) $s \to X(s,x)$ is constant during each excursion above x.

Then X is identifiable.

PROOF. (i) and (ii) assure us that $X \circ \Gamma^{-1}(t,x) = X(\rho(t,x), x)$ is continuous. This plus (ii) implies, just as in the one-parameter case, that $X \circ \Gamma^{-1}$ is predictable. It then follows from Corollary A3 that X is $\mathcal{I}$-measurable. *Q.E.D.*

COROLLARY A6. Let ϕ be identifiable. Then for each x, $\sup_t \phi(t,x)$ is $\mathcal{E}_x$-measurable.

PROOF. There exists a $\mathcal{P}$-measurable ψ such that $\phi = \psi \circ \Gamma$ and $\psi = \phi \circ \Gamma^{-1}$. Thus $\sup_t \phi(t,x) = \sup_t \psi(t,x)$. Since $\mathcal{E}_x$ is complete and $t \to \phi(t,x)$ is Borel measurable and $\mathcal{E}_x$-adapted, $\sup_t \psi(t,x)$ is $\mathcal{E}_x$-measurable. *Q.E.D.*

References

[1] R. CAIROLI and J. B. WALSH. Stochastic integrals in the plane. *Acta Math.*, *134* (1975), 111-183.

[2] E. PERKINS. Local times and semi-martingales (Preprint).

[3] J. B. WALSH. Excursions and local time. *Astérisque 52 - 53* (1978), 159 - 192.

[4] D. WILLIAMS. Conditional excursion theory. *Séminaire de Probabilités XIII (Univ. Strasbourg)*, pp. 490-494. Lecture Notes in Math 721, Springer-Verlag, Berlin, 1979.

JOHN B. WALSH
Mathematics Department
University of British Columbia
Vancouver, B.C. V6T 1W5 CANADA

PROGRESS IN PROBABILITY AND STATISTICS

Already published

PPS 1 Seminar on Stochastic Processes, 1981
E. Cınlar, K.L. Chung, R.K. Getoor, editors
ISBN 3-7643-3072-4, 248 pages, hardcover

PPS 2 Percolation Theory for Mathematicians
Harry Kesten
ISBN 3-7643-3107-0, 432 pages, hardcover

PPS 3 Branching Processes
S. Asmussen, H. Hering
ISBN 3-7643-3122-4, 472 pages, hardcover

PPS 4 Introduction to Stochastic Integration
K.L. Chung, R.J. Williams
ISBN 0-8176-3117-8
ISBN 3-7643-3117-8, 204 pages, hardcover